Distributed Cooperative Control and Optimization for Multi-Agent Systems

Qing Wang · Bin Xin · Jie Chen

Distributed Cooperative Control and Optimization for Multi-Agent Systems

北京理工大学出版社
BEIJING INSTITUTE OF TECHNOLOGY PRESS

Springer

Qing Wang
School of Automation
Beijing Institute of Technology
Beijing, China

Bin Xin
School of Automation
Beijing Institute of Technology
Beijing, China

Jie Chen
Tongji University
Shanghai, China

ISBN 978-981-96-0949-9 ISBN 978-981-96-0950-5 (eBook)
https://doi.org/10.1007/978-981-96-0950-5

Jointly published with Beijing Institute of Technology Press
The print edition is not for sale in China (Mainland). Customers from China (Mainland) please order the print book from: Beijing Institute of Technology Press.

This Springer imprint is published by the registered company Springer Nature Singapore Pte Ltd.
The registered company address is: 152 Beach Road, #21-01/04 Gateway East, Singapore 189721, Singapore

Preface

Recently, with the rapid development of computing and communication technology, control science and technology is faced with new challenges and reforms. Multi-Agent System (MAS) is a promising area in the field of distributed artificial intelligence (DAI), as well as in the mainstream control science. The control and optimization problems of multi-agent system such as multi-robot system, multi-drone system plays an important role in industrial production and national safety, which have attracted much attention. The content of this book follows the trend of control and optimization theory and technology for multi-agent system at home and abroad and has broad application in industrial production, service industry, agriculture, and other fields, which will also promote the relative study.

The primary motivation of distributed cooperative control and optimization for multi-agent systems is defined to be multiple dynamic agents that share information or tasks to accomplish a common objective under some constraints or not. Examples might include robots operating within a manufacturing cell, unmanned aircraft in search and rescue operations or military surveillance and attack missions, arrays of micro-satellites that form a distributed large aperture radar, employees operating within an organization, and software agents. In most physical systems, the dynamic models of the agents are nonlinear and coupling. The external environment of the task executing is common complex, such as uncertainties and disturbances. It has been the focus of the scholars to propose a series of complete theories and methods to overcome the properties of nonlinearities, coupling, and disturbances.

The book gradually introduces the single-integral, second-order, high-order multi-agent systems from shallow to deep. In Chap. 1, the background knowledge is introduced, including the research background and mathematical preliminaries. In Chap. 2, single- and second-order multiagent systems with mismatched uncertainties are investigated with sliding mode control methods. On the basis of Chap. 2, state containment control for a class of nonlinear multi-agent systems with time delay is studied based on observer methods in Chap. 3. Furthermore, in Chap. 4, output containment control for nonlinear multi-agents system are developed with robust compensation method. For distributed multi-agent systems, it is vital to reduce communication and computation cost. Therefore, event-based cosensus control is investigated for nonlinear multi-agent systems in Chap. 5. Distributed optimization is an important branch of cooperative control for multi-agent system. Convex optimization problems have many important properties, including a powerful duality theory and the property that any local minimum is also a global minimum. Nonsmooth optimization refers to minimization of functions that are not necessarily convex, usually locally Lipschitz, and typically not differentiable at their minimizers. In Chap. 6, distributed convex nonsmooth optimization for multi-agent system based on proximal operator is studied. On the basis of Chap. 6, distributed proximal-gradient algorithms for nonsmooth convex optimization of second-order multi-agent systems are developed in Chap. 7. In Chap. 8, event-based optimization for Euler-Lagrangian system is studied. The most important reason for singling out the study of EL systems is that they capture a large class of contemporary engineering problems, especially some which are intractable with linear control tools.

Beijing, China — Qing Wang
Beijing, China — Bin Xin
Shanghai, China — Jie Chen

Contents

Chapter 1
Introduction

1.1 Research Background

1.1.1 Distributed Cooperative Control and Optimization

In the nature, it is a natural phenomenon that a large number of individuals gather together to form a coordinated, orderly and shocking movement scene, such as a flock of birds flying collectively in the sky, a school of fish swimming in the sea, and a pack of wolves hunting cooperatively on the land. The distribution, coordination, self-organization, stability, intelligent emergence and other characteristics of these group phenomena have attracted the scholars' interest. In order to meet the needs of engineering, Prof. Minksy from Massachusetts Institute of Technology proposed the concept of "agent" and introduced the concept of individual social behavior in biology into the field of computer science.

Multi-Agent System (MAS) is an important branch of distributed artificial intelligence, and has been an international frontier subject of artificial intelligence since the end of 20th century. It is composed of several agents with certain sensing, computing, execution and communication capabilities, which exchanges information with each other through various types of communication or sensor networks and cooperates with each other to complete the given task. Compared with the traditional single complex agent, multi-agent cooperative control can work together to improve the task efficiency and reduce consumption, such as formation control, tracking control and containment control etc. In military, multi-UAVs formation means that many UAVs fly in a special flight mode throughout the entire task completion. The aim of the formation flight is to make each UAV maintain the relative position and attitude, thus expanding the vision field and achieving the full ground or controlling a target. In civil field, the multi-robot systems can be applied to the search and rescue operation to improve the rescue efficiency. During the past 20 years, with the rapid development of computer technology, communication technology, transmission, multi-agent cooperative control has attracted more and more attention and plays an increasingly

Q. Wang et al., *Distributed Cooperative Control and Optimization for Multi-Agent Systems*, https://doi.org/10.1007/978-981-96-0950-5_1

important part in people's livelihood. As a result, distributed algorithms over multi-agent networks have attracted a significant amount of research attention due to their broad range of applications in nature science, social science and engineering.

Recent years, the interest in cooperative control of multi-agent systems has been growing greatly among researchers. In view of its broad application, the fields such as physics, sociology, biology, artificial intelligence, sensor networks and control engineering have been developed rapidly. Consensus is the fundamental problem of the cooperative control, which aims to design control laws to make the certain variables of concern reach an agreement, such as Olfati-Saber and Murray (2004), Ren and Atkins (2007), Lin and Jia (2009), Li and Zhang (2010), Cheng et al. (2016), Liu et al. (2017), Wang et al. (2018b). With further study of cooperative problems, the research scope is greatly widened including tracking (Hong et al. 2006; Li et al. 2013; Xu et al. 2014; Cheng et al. 2010), formation (Liu and Tian 2009; Dong and Hu 2016), containment control (Wang et al. 2014; Sun et al. 2017), flocking (Olfati-Saber 2006; Yu et al. 2010), etc.

Optimization is a mature and very successful area of applied mathematics. It deals with the analysis and solution of problems of finding the best element in a set. In recent years, theoretical and practical researchers on distributed optimization are spending more and more efforts to meet the challenge of rapid development of high techniques, particularly the new areas such as the cloud computation and big data sets. Moreover, distributed optimization is increasingly applied to many fields such as traffic engineering, mobile robots cooperative transportation Chen and Kai (2018), smart grids Zhang and Chow (2012) and so on. Specially, distributed optimization for multi-agent systems has been widely developed in science and engineering such as distributed optimal consensus (Wang et al. 2018a; Zhao et al. 2017), resource allocation (Deng et al. 2018), and distributed optimization localization (Zhang et al. 2015), just to name a few. Distributed optimization considers effective coordination between multiple agents, and can be applied to many large-scale complicated optimization problems that are difficult to be solved by centralized algorithms. How to design an effective distributed optimization algorithm with convergence and limited complexity is one of the most important objectives in optimization research. Unlike the traditional centralized strategies such as competitive mechanism integrated whale optimization algorithms (Zeng et al. 2021) and heuristic algorithms such as particle swarm optimization approaches (Zeng et al. 2022; Liu et al. 2019c), the distributed framework has better performance (Xin et al. 2020), such as stable robustness rather than needing malfunction detection, lower communication bandwidth requirement and computation load, and higher extendibility and elasticity, which already have made significant applications on distributed algorithms to deal with optimization problems (Boyd and Vandenberghe 2004; Chen and Sayed 2012).

The design of distributed algorithms has been a frontier research topic in multi-agent analysis and control. In this decade, distributed algorithms for optimization have been extensively studied due to its wide applications. This book applies the distributed optimization approach to solve distributed optimization with equations/inequalities constraints. Furthermore, in order to reduce communication and computation cost, event-triggered scheme is adopted to solve the distributed optimization.

1.1.2 Distributed Constrained Convex Optimization

Distributed optimization, formulated in Wang and Elia (2010), Nedic et al. (2010), seeks a global optimal solution to the objective function as a sum of the local objective functions, which can be divided into unconstrained optimization and constrained optimization to fulfill some special demands such as security, privacy and kinetics. The constrained optimization problems comprise of a desirable global objective function subject to bound, equality and inequality constrains. By considering different types of constraints, including local constraints and coupled constraints, distributed optimization problems has become increasingly popular. One widely investigated class of distributed optimization problems is problems with consensus constraints and various local constraints, which cover the famous consensus problem. Distributed algorithms for optimization method any set constraint were studied recently (Gharesifard and Cortes 2014; Shi et al. 2013). Considering distributed optimization problems with consensus constraints under different conditions, both discrete-time distributed optimization algorithms (e.g., Lobel and Ozdaglar 2010; Nedic et al. 2010) and continuous-time multi-agent solvers (e.g., Wang and Elia 2010; Gharesifard and Cortes 2014; Kia et al. 2015; Qiu et al. 2016; Shi et al. 2013; Yi et al. 2015) have been developed for distributed optimization problems. Another class of popular distributed optimization problems, which is also called resource allocation, is with coupled global equality or inequality constraints. Considering global coupled constraints under various conditions, Le et al. (2019), Li et al. (2018b), Liang et al. (2018), Yi et al. (2015, 2016), Zhu et al. (2021) have studied distributed resource allocation problems, which have many practical applications such as networked control systems, traffic flow optimization, sensor fusion, and power grid control.

One widely used framework for distributed optimization is the projected primal-dual method or the method of Lagrange multipliers. To deal with set constraints, the projection method has been widely adopted in the algorithm design for constrained optimization (Ruszczyski 2006) and related problems (Nagurney and Zhang 1995). Nagurney and Zhang (1995), Yi et al. (2015), Zeng et al. (2017) constructed primal-dual type continuous-time projected algorithms to solve a distributed optimization problem, where each agent has its own private constraint function. Besides, primal-dual methods have been extended to nonsmooth objective functions with various constraints from various viewpoints in Gharesifard and Cortes (2014), Nedic et al. (2010), Liu and Wang (2015), Zeng et al. (2017). Subgradients and projections have been widely found for distributed nonsmooth optimization problems with constraints (Liu and Wang 2013; Nedic et al. 2010), and moreover, novel distributed algorithms have been proposed in the form of differential inclusions to solve such problems in the literature (see Liu and Wang 2013).

1.1.3 Distributed Optimization Based on Event-Triggered Control

In practical systems, limited energy and communication bandwidth are inevitable. It is meaningful to propose event-triggered strategies to avoid continuous communication among agents (Dimarogonas et al. 2012; Liu et al. 2018, 2019a, d; Tran et al. 2019; Wu et al. 2018; Zhang et al. 2021). Up to now, the majority of event-triggered optimal problems have been involved in single- and double-integrator dynamics (Deng et al. 2017; Li et al. 2018a; Shi et al. 2018; Yi et al. 2018; Zhao et al. 2018). In Yi et al. (2018), a distributed algorithm based on event-triggered scheme for double-integrator multi-agent systems was developed to achieve distributed optimization. In Deng et al. (2017), a distributed optimization algorithm was proposed for continuous-time multi-agent systems with discrete-time communication and gradient measurement. In Li et al. (2018a), a quantized event-triggered algorithm was considered to solve a large class of convex optimization problems for a set of single-integrator agents. In Zhao et al. (2018), a distributed event-triggered algorithm was developed to solve the convex quadratic optimization problem for a group of single-integrator multi-agent systems under undirected and connected topologies. In Shi et al. (2018), a distributed sub-gradient algorithm based on event-triggered scheme was developed for the resource allocation problem (RAP) with constraints under fixed directed topology.

There are a few results about optimal coordination of EL systems with event-triggered schemes such as Liu et al. (2019b) and Wang et al. (2017). In Liu et al. (2019b), three different distributed event-triggered control algorithms were developed to achieve leader-follower consensus for the EL multi-agent systems, but continuous communication was still inevitable when the triggering condition was monitored. In Wang et al. (2017), an event-based continuous-time algorithm for the EL system was developed to achieve distributed optimal consensus when the model parameters of the EL system were available. However, there are some limitations that the algorithms monitor the triggering function continuously or rely on the model parameters of the EL system. On one hand, the effectiveness of the event-triggered scheme can be improved by monitoring the triggering function discretely. On the other hand, it is usually difficult to obtain the model parameters exactly due to the uncertainties, so it is practical to study the optimization problem for EL systems with unavailable model parameters.

1.2 Mathematical Preliminaries

In this section, we introduce relevant mathematical concepts, and preliminaries for this report.

1.2.1 Convex Analysis

Convexity plays an important role in the study of optimization problems. We introduce convex definitions and Lipschitz continuity of functions as follows.

Definition 1.1 (*Convex function*) Ozdaglar (2003), Zhang and Hong (2015) The function $f(\cdot) : S \to \mathbb{R}$ is called convex if for any $0 \leq \mu \leq 1$

$$f(\mu x + (1-\mu)y) \leq \mu f(x) + (1-\mu) f(y), \quad \forall x, y \in S,$$

is satisfied, where S is a convex set.

If a differentiable function $f(\cdot)$ is convex over S, it has

$$f(x) - f(y) \geq \nabla f(y)^T (x-y), \quad \forall x, y \in S,$$

and $f(\cdot)$ is strictly convex if

$$f(x) - f(y) > \nabla f(y)^T (x-y),$$

whenever $x \neq y$. Furthermore, $f(\cdot)$ is said to be m-strongly convex ($m > 0$) if

$$(\nabla f(x) - \nabla f(y))^T (x-y) \geq m\|x-y\|^2, \quad \forall x, y \in S,$$

is satisfied, where $\nabla f(x)$ is the gradient of $f(\cdot)$ at x.

Definition 1.2 (*θ-Lipschitz*) Hui et al. (2009), Wei et al. (2020) A function $f(\cdot) : S \to \mathbb{R}^n$ is Lipschitz with constant $\theta > 0$, or simply θ-Lipschitz over the set S if

$$\|f(x) - f(y)\| \leq \theta \|x-y\|, \quad \forall x, y \in S,$$

is satisfied.

1.2.2 Proximal Operator

Let $g(\cdot)$ be a lower semi-continuous convex function. The proximal operator of $g(\cdot)$ is Parikh and Stephen (2014)

$$\text{prox}_g(v) = \arg\min_x \{g(x) + \frac{1}{2}\|x-v\|^2\}.$$

Define the indicator function of a closed convex set Ω as $I_\Omega(x) = 0$ if $x \in \Omega$ and $I_\Omega(x) = +\infty$ otherwise. We have $\text{prox}_{I_\Omega}(v) = P_\Omega(v)$, where $P_\Omega(v) = \arg\min_{x\in\Omega} \|x-v\|$ is the projection operator.

Let $\partial g(x)$ denote the subdifferential of $g(\cdot)$ at x. Then $\partial g(x)$ is monotone, that is,

$$(p_x - p_y)^T (x - y) \geq 0$$

for all x, y, where $p_x \in \partial g(x)$, and $p_y \in \partial g(y)$. $x = \text{prox}_g(v)$ is equivalent to

$$v - x \in \partial g(x). \tag{1.1}$$

1.2.3 Convergence Property

Consider a system

$$\dot{x}(t) = \phi(x(t)), \quad x(0) = x_0, \quad t \geq 0, \tag{1.2}$$

where $\phi : \mathbb{R}^q \to \mathbb{R}^q$ is Lipschitz continuous.

The following result is a version of the invariance principle Hui et al. (2009).

Lemma 1.1 *Let $\mathcal{D}$ be a compact, positive invariant set with respect to system (1.2), $V : \mathbb{R}^q \to \mathbb{R}$ be a continuously differentiable function, and $x(\cdot)$ be a solution of (1.2) with $x(0) = x_0 \in \mathcal{D}$. Assume $\dot{V}(x) \leq 0, \ \forall x \in \mathcal{D}$. Define $\mathcal{Z} = \{x \in \mathcal{D} : \dot{V}(x) = 0\}$, and $\mathcal{M}$ is the largest invariant subset of $\overline{\mathcal{Z}} \bigcap \mathcal{D}$, where $\overline{\mathcal{Z}}$ is the closure of $\mathcal{Z} \subset \mathbb{R}^q$. If $\dot{V}(x) \leq 0$ for all $x \in \mathcal{D}$, then* $\text{dist}(x(t), \mathcal{M}) \to 0$ *as $t \to \infty$.*

1.2.4 Optimality Condition for Convex Optimization

Consider an optimization problem

$$\min_{x \in \Omega} \ f(x) \quad \text{s.t. } g_i(x) \leq 0, \ h_j(x) = 0, \tag{1.3}$$

where $i = 1, \ldots, m, \ j = 1, \ldots, q, \ f(\cdot)$ and $g_i(\cdot)$'s are convex functions for $i \in \{1, \ldots, m\}$, $h_j(\cdot)$'s are affine functions for $j \in \{1, \ldots, q\}$, and Ω is a convex and closed set. *Slater's condition* is satisfied if there exists $x \in \text{int}(\Omega)$ such that $g_i(x) < 0$, $i = 1, \ldots, m$, and $h_j(x) = 0, \ j = 1, \ldots, q$. The following lemma introduces the Karush-Kuhn-Tucker (KKT) optimality condition (Ruszczyski 2006, Theorem 3.34).

Lemma 1.2 *Consider problem (1.3) and assume Slater's condition is satisfied. A point x^* is a solution to problem (1.3) if and only if there exist $\lambda_i^* \geq 0, \ i = 1, \ldots, m$, and $\mu_j^* \in \mathbb{R}, \ j = 1, \ldots, q$, such that $g_i(x^*) \leq 0, \ h_j(x^*) = 0, \ \lambda_i^* g_i(x^*) = 0$, and*

$$x^* \in P_\Omega \left[x^* - \partial f(x^*) - \sum_{i=1}^{m} \lambda_i^* \partial g_i(x^*) - \sum_{j=1}^{q} \mu_i^* \nabla h_j(x^*) \right].$$

1.2.5 Differential Equation

Consider a dynamical system

$$\dot{x}(t) = \phi(x(t)), \quad x(0) = x_0, \quad t \geq 0, \tag{1.4}$$

where $\phi : \mathbb{R}^q \to \mathbb{R}^q$ is Lipschitz continuous. Given a trajectory $x : [0, \infty) \to \mathbb{R}^q$ of (1.4), y is a positive limit point of $x(\cdot)$ if there is a positive increasing divergent sequence $\{t_i\}_{i=1}^{\infty} \subset \mathbb{R}$ such that $y = \lim_{i\to\infty} x(t_i)$, and a positive limit set of $x(\cdot)$ is the set of all positive limit points of $x(\cdot)$. A set $\mathcal{D}$ is said to be *positive invariant* with respect to (1.4) if $x(t) \in \mathcal{D}$ for all $t \geq 0$ and every $x_0 \in \mathcal{D}$.

Denote $\mathcal{B}_\epsilon(x)$, $x \in \mathbb{R}^n$ with a constant $\epsilon > 0$ as the open ball *centered* at x with *radius* ϵ. Let $\mathcal{D} \subset \mathbb{R}^q$ be a positive invariant set with respect to (1.4) and $z \in \mathcal{D}$ be an equilibrium of (1.4). z is *Lyapunov stable* if, for every $\epsilon > 0$, there exists $\delta = \delta(\epsilon) > 0$ such that, for every initial condition $x_0 \in \mathcal{B}_\delta(z) \bigcap \mathcal{D}$, the solution $x(t)$ of (1.4) stays in $\mathcal{B}_\epsilon(z)$ for all $t \geq 0$.

The following Lemmas are needed in the analysis of this paper.

Lemma 1.3 (Hui et al. 2009, Theorem 3.1) *Let $\mathcal{D}$ be a compact, positive invariant set with respect to system (1.4), $V : \mathbb{R}^q \to \mathbb{R}$ be a continuously differentiable function, and $x(\cdot) \in \mathbb{R}^q$ be a solution of (1.4) with $x(0) = x_0 \in \mathcal{D}$. Assume*

$$\dot{V}(x) \leq 0, \quad \forall x \in \mathcal{D},$$

and define $\mathcal{Z} = \{x \in \mathcal{D} : \dot{V}(x) = 0\}$. If every point in the largest invariant subset $\mathcal{M}$ of $\overline{\mathcal{Z}} \bigcap \mathcal{D}$ is Lyapunov stable, where $\overline{\mathcal{Z}}$ is the closure of $\mathcal{Z} \subset \mathbb{R}^n$, then (1.4) converges to one of its Lyapunov stable equilibria for any $x_0 \in \mathcal{D}$.

Lemma 1.4 *Suppose $\phi(x) = Mx + b$ with $M \in \mathbb{R}^{q\times q}$, $b \in \mathbb{R}^q$, and $\mathcal{D} = \mathbb{R}^q$. The following statements are equivalent.*

(i) System (1.4) converges to an equilibrium exponentially for any initial condition.
(ii) System (1.4) converges to an equilibrium for any initial condition.

Proof It is trivial that (i)⇒(ii) and the proof is omitted.

Suppose (ii) holds. Let x^* be an equilibrium of (1.4) and define $y = x - x^*$. System (1.4) is equivalent to

$$\dot{y}(t) = My(t) \quad y(0) = y_0 \in \mathbb{R}^q \quad t \geq 0. \tag{1.5}$$

To show (ii)⇒(i), we show that $y(t)$ converges to an equilibrium exponentially for any initial condition.

It follows from statement (ii) and Definition 11.8.1 of Bernstein (2009, p. 727) that M is semistable (that is, its eigenvalues lie on the open left half complex plane, except for a few semi-simple zero eigenvalues). Hence, there exists an invertible matrix $P \in \mathbb{R}^{q\times q}$ such that $PMP^{-1} = \begin{bmatrix} D & 0_{r\times(q-r)} \\ 0_{(q-r)\times r} & 0_{(q-r)\times(q-r)} \end{bmatrix}$, where $D \in \mathbb{R}^{r\times r}$ is Hurwitz and $r = \operatorname{rank} M \leq q$. Define $z = \begin{bmatrix} z_1 \\ z_2 \end{bmatrix} = Py \in \mathbb{R}^q$ such that $z_1 \in \mathbb{R}^r$, and $z_2 \in \mathbb{R}^{q-r}$. It follows from (1.5) that

$$\dot{z}_1(t) = Dz_1(t) \quad \dot{z}_2(t) = 0_{q-r} \quad z_1(0) = z_{1,0} \quad z_2(0) = z_{2,0} \quad t \geq 0.$$

Hence, $\left\| z(t) - \begin{bmatrix} 0_r \\ z_2(0) \end{bmatrix} \right\| = \|z_1(t)\| = \|e^{Dt}z_1(0)\|$. Recall that D is Hurwitz. The trajectory $z(t)$ converges to $\begin{bmatrix} 0_r \\ z_2(0) \end{bmatrix}$ exponentially and, equivalently, $y(t)$ converges to $P^{-1}\begin{bmatrix} 0_r \\ z_2(0) \end{bmatrix}$ exponentially.

1.2.6 Differential Inclusion

Following Aubin and Cellina (1984), a differential inclusion is given by

$$\dot{x}(t) \in \mathcal{F}(x(t)), \quad x(0) = x_0, \quad t \geq 0, \tag{1.6}$$

where $\mathcal{F}$ is a set-valued map from points in $\mathbb{R}^q$ to the nonempty, compact, convex subsets of $\mathbb{R}^q$. For each state $x \in \mathbb{R}^q$, system (1.6) specifies a *set* of possible evolutions rather than a single one. A solution of (1.6) defined on $[0, \tau] \subset [0, \infty)$ is an absolutely continuous function $x : [0, \tau] \to \mathbb{R}^q$ such that (1.6) holds for almost all $t \in [0, \tau]$ for $\tau > 0$. The solution $t \mapsto x(t)$ to (1.6) is a *right maximal solution* if it cannot be extended forward in time. Suppose that all right maximal solutions to (1.6) exist on $[0, \infty)$. A set $\mathcal{M}$ is said to be *weakly invariant* (resp., *strongly invariant*) with respect to (1.6) if, for every $x_0 \in \mathcal{M}$, $\mathcal{M}$ contains a maximal solution (resp., all maximal solutions) of (1.6). A point $z \in \mathbb{R}^q$ is a *positive limit point* of a solution $\phi(t)$ to (1.6) with $\phi(0) = x_0 \in \mathbb{R}^q$, if there exists a sequence $\{t_k\}_{k=1}^{\infty}$ with $t_k \to \infty$ and $\phi(t_k) \to z$ as $k \to \infty$. The set $\omega(\phi(\cdot))$ of all such positive limit points is the *positive limit set* for the trajectory $\phi(t)$ with $\phi(0) = x_0 \in \mathbb{R}^q$.

An equilibrium point of (1.6) is a point $x_e \in \mathbb{R}^q$ such that $0_q \in \mathcal{F}(x_e)$. It is easy to see that x_e is an equilibrium point of (1.6) if and only if the constant function $x(\cdot) = x_e$ is a solution of (1.6). An equilibrium point $z \in \mathbb{R}^q$ of (1) is *Lyapunov stable* if, for every $\varepsilon > 0$, there exists $\delta = \delta(\varepsilon) > 0$ such that, for every initial condition $x(0) = x_0 \in \mathcal{B}_\delta(z)$, every solution $x(t) \in \mathcal{B}_\varepsilon(z)$ for all $t \geq 0$.

Let $V : \mathbb{R}^q \to \mathbb{R}$ be a locally Lipschitz continuous convex function, and let ∂V be the *subgradient* of $V(x)$ at x. The *set-valued Lie derivative* $\mathcal{L}_{\mathcal{F}}V : \mathbb{R}^q \to \mathcal{P}(\mathbb{R})$ of V with respect to (1.6) is defined as $\mathcal{L}_{\mathcal{F}}V(x) = \{a \in \mathbb{R} :$ there exists $v \in \mathcal{F}(x)$ such that $p^{\mathrm{T}}v = a$ for all $p \in \partial V(x)\}$. More notations and definitions about nonsmooth analysis can be found in Aubin and Cellina (1984). In the case where $\mathcal{L}_{\mathcal{F}}V(x)$ is nonempty, we use $\max \mathcal{L}_{\mathcal{F}}V(x)$ to denote the largest element of $\mathcal{L}_{\mathcal{F}}V(x)$. Recall that if $\phi(\cdot)$ is a solution of (1.6) and $V : \mathbb{R}^q \to \mathbb{R}$ is locally Lipschitz and *regular* (see (Bacciotti and Ceragioli, 1999, p. 4) or (Clarke, 1990, p. 39)), then $\dot{V}(\phi(t))$ exists almost everywhere, and $\dot{V}(\phi(t)) \in \mathcal{L}_{\mathcal{F}}V(\phi(t))$ almost everywhere. Recall that the solution $t \mapsto x(t)$ to (1.6) is a *right maximal solution* if it cannot be extended forward in time. We assume that all right maximal solutions to (1.6) exist on $[0, \infty)$. A set $\mathcal{M}$ is said to be *weakly invariant* (Bacciotti and Ceragioli 1999; Ryan 1998) (resp., *strongly invariant*) with respect to (1.6) if for every $x_0 \in \mathcal{M}$, $\mathcal{M}$ contains a maximal solution (Bacciotti and Ceragioli 1999; Ryan 1998) (resp., all maximal solutions) of (1.6).

A set-valued map $\mathcal{F} : \mathbb{R}^q \to \mathcal{P}(\mathbb{R}^q)$ is called upper semi-continuous (see Aubin and Cellina 1984) at x if for each open set $V \supseteq \mathcal{F}(x)$, there exists a neighborhood U of x such that $\mathcal{F}(y) \subset V$ for all $x \in U$. Recall that if $\mathcal{F}$ is locally bounded and upper semicontinuous with nonempty, compact, and convex values, then there exist solutions for (1.6) (see Aubin and Cellina 1984). A set-valued map $\mathcal{F} : \mathbb{R}^q \to \mathfrak{B}(\mathbb{R}^q)$ is said to be monotone on $\Omega \subset \mathbb{R}^q$ if for all $f_1 \in \mathcal{F}(x_1)$, $f_2 \in \mathcal{F}(x_2)$, and for all $x_1, x_2 \in \Omega$,

$$\langle x_1 - x_2, f_1 - f_2 \rangle \geq 0, \tag{1.7}$$

where $\langle \cdot, \cdot \rangle$ denotes the usual inner product in $\mathbb{R}^q$. In addition, $\mathcal{F}$ is strictly monotone if the inequality in (1.7) holds whenever $x_1 \neq x_2$.

Next, we introduce a version of the invariance principle (Theorem 2 of Cortes 2008), which is based on nonsmooth regular functions.

Lemma 1.5 *For the differential inclusion (1.6), we assume that $\mathcal{F}$ is upper semicontinuous and locally bounded, and $\mathcal{F}(x)$ takes nonempty, compact, and convex values. Let $V : \mathbb{R}^q \to \mathbb{R}$ be a locally Lipschitz and regular function, $\mathcal{S} \subset \mathbb{R}^q$ be compact and strongly invariant for (1.6), $\phi(\cdot)$ be a solution of (1.6),*

$$\mathcal{R} = \{x \in \mathbb{R}^q : 0 \in \mathcal{L}_{\mathcal{F}}V(x)\},$$

and $\mathcal{M}$ be the largest weakly invariant subset of $\overline{\mathcal{R}} \cap \mathcal{S}$, where $\overline{\mathcal{R}}$ is the closure of $\mathcal{R}$. If $\max \mathcal{L}_{\mathcal{F}}V(x) \leq 0$ *for all $x \in \mathcal{S}$, then* $\mathrm{dist}(\phi(t), \mathcal{M}) \to 0$ *as $t \to +\infty$.*

1.3 Brief Outline of the Book

The book gradually introduces the single-integral, second-order, high-order multi-agent systems from shallow to deep. The remainder of the book is organized as follows. In this chapter, the background knowledge is introduced, including the research background and mathematical preliminaries. In Chap. 2, single- and second-order multi-agent systems with mismatched uncertainties are investigated with sliding mode control methods. On the basis of Chap. 2, state containment control for a class of nonlinear multi-agent systems with time delay is studied based on observer methods in Chap. 3. Furthermore, in Chap. 4, output containment control for nonlinear multi-agents system are developed with robust compensation method. For distributed multi-agent systems, it is vital to reduce communication and computation cost. Therefore, event-based cosensus control is investigated for nonlinear multi-agent systems in Chap. 5. Distributed optimization is an important branch of cooperative control for multi-agent system. In Chap. 6, distributed convex nonsmooth optimization for multi-agent system based on proximal operator is studied. On the basis of Chap. 6, distributed proximal-gradient algorithms for nonsmooth convex optimization of second-order multi-agent systems are developed in Chap. 7. In Chap. 8, event-based optimization for Euler-Lagrangian system is studied.

References

Aubin, J.-P. and Cellina, A. (1984). *Differential Inclusions* (Springer, Berlin, Germany).

Bacciotti, A. and Ceragioli, F. (1999). Stability and stabilization of discontinuous systems and nonsmooth lyapunov functions, *ESAIM: Control, Optimisation and Calculus of Variations*4, pp. 361–376.

Bernstein, D. S. (2009). *Matrix mathematics: theory, facts, and formulas* (Princeton university press).

Boyd, S. and Vandenberghe, L. (2004). *Convex Optimization.*

Chen, J., and Sayed, A.H. (2012). Diffusion adaptation strategies for distributed optimization and learning over networks, *IEEE Transactions on Signal Processing*60, 8, pp. 4289–4305.

Chen, J. and Kai, S. X. (2018). Cooperative transportation control of multiple mobile manipulators through distributed optimization, *Science China Information Sciences*61, 12.

Cheng, L., Hou, Z.-G., Tan, M., Lin, Y., and Zhang, W. (2010). Neural-network-based adaptive leader-following control for multiagent systems with uncertainties, *IEEE Transactions on Neural Networks*21, 8, pp. 1351–1358.

Cheng, L., Wang, H., Hou, Z.-G., and Tan, M. (2016). Reaching a consensus in networks of high-order integral agents under switching directed topologies, *International Journal of Systems Science*47, 8, pp. 1966–1981.

Clarke, F. H. (1990). *Optimization and nonsmooth analysis* (SIAM).

Cortes, J. (2008). Discontinuous dynamical systems, *IEEE Control systems magazine*28, 3, pp. 36–73.

Deng, Z. H., Wang, X. H., and Hong, Y. G. (2017). Distributed optimisation design with triggers for disturbed continuous-time multi-agent systems, *IET Control Theory and Applications*11, 2, pp. 282–290.

Deng, Z. H., Liang, S., and Yu, W. Y. (2018). Distributed optimal resource allocation of second-order multiagent systems, *International Journal of Robust and Nonlinear Control*28, 14, pp. 4246–4260.

Dimarogonas, D. V., Frazzoli, E., and Johansson, K. H. (2012). Distributed event-triggered control for multi-agent systems, *IEEE Transactions on Automatic Control*57, 5, pp. 1291–1297.

Dong, X. and Hu, G. (2016). Time-varying formation control for general linear multi-agent systems with switching directed topologies, *Automatica*73, pp. 47–55.

Gharesifard, B. and Cortes, J. (2014). Distributed continuous-time convex optimization on weight-balanced digraphs, *IEEE Transactions on Automatic Control*59, 3, pp. 781–786.

Hong, Y., Hu, J., and Gao, L. (2006). Tracking control for multi-agent consensus with an active leader and variable topology, *Automatica*42, 7, pp. 1177–1182.

Hui, Q., Haddad, W. M., and Bhat, S. P. (2009). Semistability, finite-time stability, differential inclusions, and discontinuous dynamical systems having a continuum of equilibria, *IEEE Transactions on Automatic Control***54**, 10, pp. 2465–2470.

Kia, S. S., Cortes, J., and Martinez, S. (2015). Distributed convex optimization via continuous-time coordination algorithms with discrete-time communication, *Automatica***55**, pp. 254–264.

Le, X., Chen, S., Li, F., Yan, Z., and Xi, J. (2019). Distributed neurodynamic optimization for energy internet management, *IEEE Transactions on Systems, Man, and Cybernetics: Systems***49**, 8, pp. 1624–1633.

Li, T. and Zhang, J.F. (2010). Consensus conditions of multi-agent systems with time-varying topologies and stochastic communication noises, *IEEE Transactions on Automatic Control***55**, 9, pp. 2043–2057.

Li, Z., Liu, X., Ren, W., and Xie, L. (2013). Distributed tracking control for linear multiagent systems with a leader of bounded unknown input, *IEEE Transactions on Automatic Control***58**, 2, pp. 518–523.

Li, H. Q., Liu, S., Soh, Y. C., and Xie, L. H. (2018a). Event-triggered communication and data rate constraint for distributed optimization of multiagent systems, *IEEE Transactions on Systems Man Cybernetics-Systems***48**, 11, pp. 1908–1919.

Li, K., Liu, Q., Yang, S., Cao, J., and Lu, G. (2018b). Cooperative optimization of dual multiagent system for optimal resource allocation, *IEEE Transactions on systems, man, and cybernetics: systems***50**, 11, pp. 4676–4687.

Liang, S., Zeng, X. L., and Hong, Y. G. (2018). Distributed nonsmooth optimization with coupled inequality constraints via modified lagrangian function, *IEEE Transactions on Automatic Control***63**, 6, pp. 1753–1759.

Lin, P. and Jia, Y. (2009). Consensus of second-order discrete-time multi-agent systems with nonuniform time-delays and dynamically changing topologies, *Automatica***45**, 9, pp. 2154–2158.

Liu, C.-L. and Tian, Y.-P. (2009). Formation control of multi-agent systems with heterogeneous communication delays, *International Journal of Systems Science***40**, 6, pp. 627–636.

Liu, Q. S. and Wang, J. (2013). A one-layer projection neural network for nonsmooth optimization subject to linear equalities and bound constraints, *IEEE Transactions on Neural Networks and Learning Systems***24**, 5, pp. 812–824.

Liu, Q. S. and Wang, J. (2015). A second-order multi-agent network for bound-constrained distributed optimization, *IEEE Transactions on Automatic Control***60**, 12, pp. 3310–3315.

Liu, J., Yu, Y., Wang, Q., and Sun, C. (2017). Fixed-time event-triggered consensus control for multi-agent systems with nonlinear uncertainties, *Neurocomputing***260**, pp. 497–504.

Liu, J., Yu, Y., Sun, J., and Sun, C. Y. (2018). Distributed event-triggered fixed-time consensus for leader-follower multiagent systems with nonlinear dynamics and uncertain disturbances, *International Journal of Robust and Nonlinear Control***28**, 11, pp. 3543–3559.

Liu, J., Zhang, Y. L., Sun, C. Y., and Yu, Y. (2019a). Fixed-time consensus of multi-agent systems with input delay and uncertain disturbances via event-triggered control, *Information Sciences***480**, pp. 261–272.

Liu, Q. C., Ye, M. B., Qin, J. H., and Yu, C. B. (2019b). Event-triggered algorithms for leader-follower consensus of networked euler-lagrange agents, *IEEE Transactions on Systems Man Cybernetics-Systems***49**, 7, pp. 1435–1447.

Liu, W., Wang, Z., Liu, X., Zeng, N., and Bell, D. (2019c). A novel particle swarm optimization approach for patient clustering from emergency departments, *IEEE Transactions on Evolutionary Computation***23**, 4, pp. 632–644.

Liu, J., Zhang, Y., Yu, Y., and Sun, C. (2019d). Fixed-time event-triggered consensus for nonlinear multiagent systems without continuous communications, *IEEE Transactions on Systems, Man, and Cybernetics: Systems***49**, 11, pp. 2221–2229.

Lobel, I. and Ozdaglar, A. (2010). Distributed subgradient methods for convex optimization over random networks, *IEEE Transactions on Automatic Control***56**, 6, pp. 1291–1306.

Nagurney, A. and Zhang, D. (1995). *Projected dynamical systems and variational inequalities with applications*, Vol. 2 (Springer Science & Business Media).

Nedic, A., Ozdaglar, A., and Parrilo, P. A. (2010). Constrained consensus and optimization in multi-agent networks, *IEEE Transactions on Automatic Control***55**, 4, pp. 922–938.
Olfati-Saber, R. and Murray, R.M. (2004). Consensus problems in networks of agents with switching topology and time-delays, *IEEE Transactions on Automatic Control***49**, 9, pp. 1520–1533.
Olfati-Saber, R. (2006). Flocking for multi-agent dynamic systems: Algorithms and theory, *IEEE Transactions on Automatic Control***51**, 3, pp. 401–420.
Ozdaglar, D. P. B. A. (2003). *Convex Analysis and Optimization* (Athena Scientific).
Parikh, N. and Stephen, B. (2014). Proximal algorithms, *Foundations and Trends extregistered in Optimization***3**, 1, pp. 127–239.
Qiu, Z. R., Liu, S., and Xie, L. H. (2016). Distributed constrained optimal consensus of multi-agent systems, *Automatica***68**, pp. 209–215.
Ren, W. and Atkins, E. (2007). Distributed multi-vehicle coordinated control via local information exchange, *International Journal of Robust and Nonlinear Control***17**, 10-11, pp. 1002–1033.
Ruszczyski, A. (2006). *Nonlinear Optimization* (Princeton University Press).
Ryan, E. (1998). An integral invariance principle for differential inclusions with applications in adaptive control, *SIAM Journal on Control and Optimization***36**, 3, pp. 960–980.
Shi, G. D., Johansson, K. H., and Hong, Y. G. (2013). Reaching an optimal consensus: Dynamical systems that compute intersections of convex sets, *IEEE Transactions on Automatic Control***58**, 3, pp. 610–622.
Shi, X. S., Wang, Y. D., Song, S. M., and Yan, G. F. (2018). Distributed optimisation for resource allocation with event-triggered communication over general directed topology, *International Journal of Systems Science***49**, 6, pp. 1119–1130.
Sun, C., Wang, Q., and Yu, Y. (2017). Robust output containment control of multi-agent systems with unknown heterogeneous nonlinear uncertainties in directed networks, *International Journal of Systems Science***48**, 6, pp. 1173–1181.
Tran, N. T., Wang, Y. W., Liu, X. K., Xiao, J. W., and Lei, Y. (2019). Distributed optimization problem for second-order multi-agent systems with event-triggered and time-triggered communication, *Journal of the Franklin Institute-Engineering and Applied Mathematics***356**, 17, pp. 10196–10215.
Wang, J. and Elia, N. (2010). Control approach to distributed optimization, in *2010 48th Annual Allerton Conference on Communication, Control, and Computing (Allerton)* (IEEE), pp. 557–561.
Wang, Y., Cheng, L., Hou, Z.-G., Tan, M., and Wang, M. (2014). Containment control of multi-agent systems in a noisy communication environment, *Automatica***50**, 7, pp. 1922–1928.
Wang, X. F., Deng, Z. H., Ma, S., and Du, X. (2017). Event-triggered design for multi-agent optimal consensus of euler-lagrangian systems, *Kybernetika***53**, 1, pp. 179–194.
Wang, A. J., Liao, X. F., and Dong, T. (2018a). Event-triggered gradient-based distributed optimisation for multi-agent systems with state consensus constraint, *IET Control Theory and Applications***12**, 10, pp. 1515–1519.
Wang, Q., Yu, Y., and Sun, C. (2018b). Distributed event-based consensus control of multi-agent system with matching nonlinear uncertainties, *Neurocomputing***272**, pp. 694–702.
Wei, Y., Fang, H., Zeng, X. L., Chen, J., and Pardalos, P. (2020). A smooth double proximal primal-dual algorithm for a class of distributed nonsmooth optimization problems, *IEEE Transactions on Automatic Control***65**, 4, pp. 1800–1806.
Wu, Z. G., Xu, Y., Lu, R. Q., Wu, Y. Q., and Huang, T. W. (2018). Event-triggered control for consensus of multiagent systems with fixed/switching topologies, *IEEE Transactions on Systems Man Cybernetics-Systems***48**, 10, pp. 1736–1746.
Xin, R., Pu, S., Nedić, A., and Khan, U.A.(2020). A general framework for decentralized optimization with first-order methods, *Proceedings of the IEEE***108**, 11, pp. 1869–1889.
Xu, W., Cao, J., Yu, W., and Lu, J. (2014). Leader-following consensus of non-linear multi-agent systems with jointly connected topology, *IET Control Theory and Applications***8**, 6, pp. 432–440.
Yi, X., Yao, L., Yang, T., George, J., and Johansson, K. H. (2018). Distributed optimization for second-order multi-agent systems with dynamic event-triggered communication, in *2018 IEEE Conference on Decision and Control (CDC)* (IEEE), pp. 3397–3402.

Yi, P., Hong, Y. G., and Liu, F. (2015). Distributed gradient algorithm for constrained optimization with application to load sharing in power systems, *Systems & Control Letters* **83**, pp. 45–52.

Yi, P., Hong, Y. G., and Liu, F. (2016). Initialization-free distributed algorithms for optimal resource allocation with feasibility constraints and application to economic dispatch of power systems, *Automatica* **74**, pp. 259–269.

Yu, W., Chen, G., and Cao, M. (2010). Distributed leader-follower flocking control for multi-agent dynamical systems with time-varying velocities, *Systems and Control Letters* **59**, 9, pp. 543–552.

Zeng, X. L., Yi, P., and Hong, Y. G. (2017). Distributed continuous-time algorithm for constrained convex optimizations via nonsmooth analysis approach, *IEEE Transactions on Automatic Control* **62**, 10, pp. 5227–5233.

Zeng, N., Song, D., Li, H., You, Y., Liu, Y., and Alsaadi, F. E. (2021). A competitive mechanism integrated multi-objective whale optimization algorithm with differential evolution q, *Neurocomputing* **432**, pp. 170–182.

Zeng, N., Wang, Z., Liu, W., Zhang, H., Hone, K., and Liu, X. (2022). A dynamic neighborhood-based switching particle swarm optimization algorithm, *IEEE Transactions on Cybernetics* **52**, 9, pp. 9290–9301.

Zhang, Z. and Chow, M.Y. (2012). Convergence analysis of the incremental cost consensus algorithm under different communication network topologies in a smart grid, *IEEE Transactions on Power Systems* **27**, 4, pp. 1761–1768.

Zhang, Y. and Hong, Y. (2015). Distributed optimization design for high-order multi-agent systems, in *2015 34th Chinese Control Conference (CCC)*, 2015 34th Chinese Control Conference (CCC), pp. 7251–7256.

Zhang, Y. Q., Lou, Y. C., Hong, Y. G., and Xie, L. H. (2015). Distributed projection-based algorithms for source localization in wireless sensor networks, *IEEE Transactions on Wireless Communications* **14**, 6, pp. 3131–3142.

Zhang, Y., Su, X. H., Liu, Z., and Chen, C. (2021). Event-triggered adaptive fuzzy tracking control with guaranteed transient performance for mimo nonlinear uncertain systems, *IEEE Transactions on Cybernetics* **51**, 2, pp. 736–749.

Zhao, Y., Liu, Y. F., Wen, G. H., and Chen, G. R. (2017). Distributed optimization for linear multiagent systems: Edge- and node-based adaptive designs, *IEEE Transactions on Automatic Control* **62**, 7, pp. 3602–3609.

Zhao, Z. Y., Chen, G., and Dai, M. X. (2018). Distributed event-triggered scheme for a convex optimization problem in multi-agent systems, *Neurocomputing* **284**, pp. 90–98.

Zhu, Y. N., Ren, W., Yu, W. W., and Wen, G. H. (2021). Distributed resource allocation over directed graphs via continuous-time algorithms, *IEEE Transactions on Systems Man Cybernetics-Systems* **51**, 2, pp. 1097–1106.

Chapter 2
Disturbance Observer-Based Sliding Mode Control for Multi-agent Systems with Mismatched Uncertainties

In practice, there exist nonlinearities and uncertainties for most multi-agent systems. For nonlinear systems, sliding mode control provides an effective way due to the advantages of robustness and simplicity. A lot of derivative methods have sprung up including feedback linearization control (Li et al. 2017), terminal sliding mode control (Yan et al. 2016; Mu et al. 2016), high order sliding mode control (Yan et al. 2016), adaptive sliding mode control and so on. In Lu et al. (2012), combined with adaptive algorithm, the sliding mode controller was designed for the attitude tracking control problem of a nonlinear spacecraft model with external disturbances and inertia uncertainties. In Li et al. (2017), a new fixed-time sliding mode control algorithm via the backstepping method was proposed for a class of high-order strict-feedback nonlinear systems (SFNSs) with mismatching system uncertainties. In Yan et al. (2016), the Euler's discretization of the second order sliding mode control system with the twisting algorithm was studied. In Yang et al. (2014), a continuous dynamic sliding-mode control method was proposed for mismatched disturbance attenuation using a high-order sliding-mode differentiator.

Due to its particular robustness to restrain disturbances and plant uncertainties, the sliding mode control is widely employed in multi-agent systems. In Ren and Chen (2015), both a new distributed asymptotic consensus controller and terminal sliding mode control were considered for the leader-following consensus problem of second-order nonlinear multi-agent systems. In Yu and Long (2015), both the discontinuous or continuous integral sliding mode (ISM) protocols were respectively developed to achieve accurate finite-time consensus in spite of the disturbances for the second-order multi-agent systems. In Han et al. (2017), the distributed finite-time formation tracking protocols were proposed via the fast terminal sliding mode control (FTSMC) scheme for the finite-time formation tracking control problem of multi-agent systems.

Chattering phenomenon is inevitable in sliding mode control. To improve the control performance, some advanced control methods such as neural networks, fuzzy control were combined with sliding mode control. In Zou et al. (2013), Chebyshev

Q. Wang et al., *Distributed Cooperative Control and Optimization for Multi-Agent Systems*, https://doi.org/10.1007/978-981-96-0950-5_2

neural networks were used in conjunction with the terminal sliding mode control. In Chang et al. (2012), a fuzzy sliding-mode formation controller (FSMFC) was proposed to address the decentralized formation problems for multiple robots. However, the chattering attenuation ability is achieved at the price of sacrificing the control performance.

Compared with the methods above, disturbance observer serves like a patch to the baseline controller and does not cause any adverse effects in the absence of uncertainties (Yang et al. 2013; Zhang et al. 2016). Besides, it can not only handle the mismatched uncertainties but has the advantages of simplicity. Therefore, the disturbance observer is introduced for sliding mode control in the chapter. In conjunction with the disturbance observer, a finite time sliding mode controller is proposed for the second order multi-agent systems with mismatched uncertainties in this chapter.

The main contributions are listed as follows: (1) The multi-agent systems described with mismatched uncertainties are considered; (2) both discontinuous and continuous sliding mode controllers are considered; furthermore, comparisons among integral, discontinuous and continuous sliding mode controllers are made to show the advantages of the proposed method; (3) with the proposed sliding mode controller, the desired sliding surface can be reached in finite time; (4) the SMC via a disturbance observer in this chapter could attenuate the mismatched disturbances without sacrificing its nominal performance and the chattering problem can be relieved to some extent.

2.1 Mathematical Preliminaries

2.1.1 Notations

In this chapter, the vector $x = [x_1, \ldots, x_N]^T$, $\alpha \in R$ and $sgn(\cdot)$ denotes the sign function. Define the function $sig^{\alpha}(x_i) = |x_i|^{\alpha} sgn(x_i)$, $i = 1, \ldots, N$, and the vector $sig^{\alpha}(x) = [sig^{\alpha}(x_1), \ldots, sig^{\alpha}(x_N)]^T$. $\mathbf{0}$ and $\mathbf{1}_N$ represent the column vectors with all elements being 0 and 1, respectively. For vectors, $\|\cdot\|_1$ and $\|\cdot\|_2$ denote the 1-norm, Euclidean norm, respectively. For instance, $\|x\|_1 = \sum_{i=1}^{N} |x_i|, \|x\|_2 = \sqrt{x^T x}$.

2.1.2 Graph Theory

For a graph $\mathcal{G}$ with n vertices, denoting the graph $\mathcal{G} = \{\mathcal{V}, \mathcal{E}\}$. $\mathcal{V} = \{v_1, v_2, \ldots, v_n\}$ represents the nodes, and $\mathcal{E} = \{(v_i, v_j)|v_i, v_j \in \mathcal{V}\}$ is the set of edges. $\mathcal{G}$ is called undirected if $(v_i, v_j) \in \mathcal{E} \Leftrightarrow (v_j, v_i) \in \mathcal{E}$. Meanwhile, if there is an edge between v_i and v_j, then it is said that node v_i and node v_j are adjacent. The adjacency matrix

$\mathcal{A} = [a_{ij}] \in R^{n \times n}$ associated with $\mathcal{G}$ is defined as $a_{ij} = 1$ if $(v_j, v_i) \in \mathcal{E}$. For each node v_i, N_i is the cardinality of its neighbor set and $N_m = \max\limits_{i=1,\ldots,n} \{N_i\}$.

2.2 Problem Formulation

Consider the second-order MAS with mismatched disturbance, depicted by

$$\begin{cases} \dot{x}_i(t) = v_i(t) + d_i(t) \\ \dot{v}_i(t) = u_i(t) \end{cases} \tag{2.1}$$

where x_i and $v_i(t)$ represent the position and velocity, respectively, $u_i(t)$ is the control input, $d_i(t)$ is the disturbance.

Assumption 2.1 For the MAS (2.1), there exists a directed spanning tree in graph $\mathcal{G}$.

Assumption 2.2 The disturbance in system (2.1) is bounded by $d^* = \sup\limits_{t>0} |d_i(t)|$ and satisfies $\lim\limits_{t \to \infty} \dot{d}_i(t) = 0$.

Lemma 2.1 *For* $x_i \in R$, $i = 1, 2, \ldots, n$, $\alpha \in (0, 1]$, *then* $(\sum\limits_{i=1}^{n} |x_i|)^\alpha \leq \sum\limits_{i=1}^{n} |x_i|^\alpha$.

Lemma 2.2 (Finite-time Lyapunov Stability Theorem) *Consider the non-Lipschiz continuous nonlinear system* $\dot{x} = f(x)$ *with* $f(0) = 0$. *Suppose there exists a continuous function* $V(x)$ *defined on a neighborhood of the origin, and real numbers* $c > 0$ *and* $0 < \alpha < 1$, *such that the following conditions hold:*

(i) $V(x)$ *is positive definite;*

(ii) $\dot{V}(x) + cV^\alpha \leq 0$.

Then the origin is locally finite-time stable, and the settling time, depending on the initial state $x(0) = x_0$, *satisfies*

$$T(X_0) \leq \frac{1}{c(1-\alpha)} V(x_0)^{1-\alpha}$$

for all x_0 *in some open neighborhood of the origin.*

2.3 Integral SMC

Integral sliding mode control is an effective method for restraining the mismatched uncertainties. Motivated by the integral sliding mode method (Sam et al. 2004), the following sliding-mode surface is selected:

$$s_i = v_i + c_1 \sum_{j=1}^{N} a_{ij}(x_i - x_j) + c_2 \sum_{j=1}^{N} \int a_{ij}(x_i - x_j) \tag{2.2}$$

The integral SMC controller is designed as

$$u_i(t) = -\left[c_1 \sum_{j=1}^{N} a_{ij}(v_i - v_j) + c_2 \sum_{j=1}^{N} a_{ij}(x_i - x_j) + ksgn(s_i) \right] \tag{2.3}$$

We can derive the derivative of s_i,

$$\begin{aligned} \dot{s}_i &= u_i + c_1 \sum_{j=1}^{N} a_{ij}(v_i - v_j) + c_1 \sum_{j=1}^{N} a_{ij}(d_i - d_j) + c_2 \sum_{j=1}^{N} a_{ij}(x_i - x_j) \\ &= -ksgn(s_i) + c_1 \sum_{j=1}^{N} a_{ij}(d_i - d_j) \end{aligned} \tag{2.4}$$

From (2.4), we can design the switching gain $k > c_1 d^*$ so that the sliding surface s_i can reach zero in finite time. Then, substituting the condition $s_i = 0$ into (2.2), yields

$$v_i = -c_1 \sum_{j=1}^{N} a_{ij}(x_i - x_j) - c_2 \sum_{j=1}^{N} \int a_{ij}(x_i - x_j) \tag{2.5}$$

Then combining (2.1) and (2.5), we can derive that

$$\ddot{x}_i + c_1 \sum_{j=1}^{N} a_{ij}(\dot{x}_i - \dot{x}_j) + c_2 \sum_{j=1}^{N} a_{ij}(x_i - x_j) = \dot{d}_i \tag{2.6}$$

Therefore, if $\lim_{t\to\infty} \dot{d}(t) = 0$, it is easy to derive that $x_i \to x_j$ and $v_i \to v_j$.

Lemma 2.3 (Ren and Atkins (2007)) *For the second-order multi-agent system*

$$\begin{cases} \dot{x}_i(t) = v_i(t) \\ \dot{v}_i(t) = u_i(t) \end{cases} \tag{2.7}$$

With the control protocol $u_i(t) = -c_1 \sum_{j=1}^{N} a_{ij}(v_i - v_j) - c_2 \sum_{j=1}^{N} a_{ij}(x_i - x_j)$, the system (2.7) can achieve consensus, and the consistent function satisfies

$$\lim_{t\to\infty} \left(x(t) - \mathbf{1}p^T x(0) - t\mathbf{1}p^T v(0) \right) = 0,$$
$$\lim_{t\to\infty} \left(v(t) - \mathbf{1}p^T v(0) \right) = 0,$$

where $\mathbf{1} = [1, 1, \ldots, 1]^T$ and p is the non-negative left eigenvector of $-L$ corresponding to eigenvalue 0 and $p^T\mathbf{1} = 1$.

Remark 2.1 If $\lim\limits_{t\to\infty} \dot{d}(t) = 0$, the equation (2.6) is equivalent to the above system (2.7), then we can have that $\lim_{t\to\infty} \left(x_i(t) - x_j(t)\right) = 0$ and $\lim_{t\to\infty} \left(v_i(t) - v_j(t)\right) = 0$.

Though the integral sliding mode method is efficient to restrain the mismatched uncertainties, it always introduces overshooting to the control performance; besides, the integral SMC doesn't do well in alleviating the chattering phenomenon.

2.4 Discontinuous SMC via a Disturbance Observer

Instead of the integral SMC, a new discontinuous sliding mode controller via a disturbance observer is proposed to alleviate the chattering phenomenon and improve the control performance in this section.

Define $z_i = [x_i, v_i]^T$, then the system (2.1) can be described as

$$\dot{z}_i = f_i(v_i) + G_1 u_i + G_2 d_i \tag{2.8}$$

where $f_i(v_i) = [v_i, 0]^T$,$G_1 = [0, 1]^T$,$G_2 = [1, 0]^T$.

Then the nonlinear disturbance observer is introduced, which is described as the following:

$$\begin{cases} \dot{p}_i = -LG_2 p_i - L[G_2 L z_i + f_i(v_i) + G_1 u_i] \\ \hat{d}_i = p_i + L z_i \end{cases} \tag{2.9}$$

where $\hat{d}_i$, p_i, and L represent the estimation of nonlinear disturbance, the internal state of the nonlinear observer and the observer gain respectively.

A novel sliding-mode surface for system (2.1) with nonlinear uncertainties is selected as the following

$$s_i = v_i + c\sum_{j=1}^{N} a_{ij}(x_i - x_j) + \hat{d}_i \tag{2.10}$$

where $\hat{d}_i$ is the estimation of disturbance d_i.

The proposed sliding mode controller based on DOB is designed as

$$u_i = -\left[c\sum_{j=1}^{N} a_{ij}(v_i - v_j) + c\sum_{j=1}^{N} a_{ij}(\hat{d}_i - \hat{d}_j) + ksgn(s_i)\right] \tag{2.11}$$

Therefore, substituting u_i into the derivation of s_i is derived,

$$\begin{aligned}\dot{s}_i &= \dot{v}_i + c\left[\sum_{j=1}^{N} a_{ij}(v_i - v_j) + \sum_{j=1}^{N} a_{ij}(\hat{d}_i - \hat{d}_j)\right] + \dot{\hat{d}}_i \\ &= -c\sum_{j=1}^{N} a_{ij}(\hat{d}_i - \hat{d}_j) + c\sum_{j=1}^{N} a_{ij}(d_i - d_j) + \dot{\hat{d}}_i - ksgn(s_i)\end{aligned} \tag{2.12}$$

According to (2.8) and (2.9), it derives that

$$\begin{aligned}\dot{\hat{d}}_i &= \dot{p}_i + L\dot{z}_i \\ &= -LG_2p_i - L[G_2Lz_i + f_i(v_i) + G_1u_i] + L[f_i(v_i) + G_1u_i + G_2d_i] \\ &= -LG_2(p_i + Lz_i) + +LG_2d_i \\ &= -LG_2\hat{d}_i + LG_2d_i\end{aligned} \tag{2.13}$$

Substituting (2.13) into (2.12), it follows that

$$\dot{s}_i = -ksgn(s_i) + Lg_2e_{d_i} + c\sum_{j=1}^{N} a_{ij}(e_{d_i} - e_{d_j}) \tag{2.14}$$

where $e_{d_i} = d_i - \hat{d}_i$ and $e_{d_j} = d_j - \hat{d}_j$.

Remark 2.2 The disturbance estimation error $e_{d_i} = d_i - \hat{d}_i$ is bounded, satisfying $e_d^* = \sup_{t>0} |e_{d_i}|, i = 1, \ldots, N$.

Proof From the definition of e_{d_i} and equation (2.13), one can derive

$$\begin{aligned}\dot{e}_{d_i} &= \dot{d}_i - \dot{\hat{d}}_i \\ &= \dot{d}_i + LG_2\hat{d}_i - LG_2d_i \\ &= -LG_2e_{d_i} + \dot{d}_i\end{aligned} \tag{2.15}$$

It can be verified that the error system (2.15) is asymptotically stable since $\lim_{t\to\infty} \dot{d}_i = 0$ is satisfied in Assumption 2.2. Therefore, the Remark 2.2 is rational.

Theorem 2.1 *Suppose that the Assumptions 2.1–2.2 hold, for the multi-agent system (2.1) and sliding-mode surface (2.10), the proposed disturbance observer (2.9) and discontinuous controller (2.11) guarantee that the sliding-mode surface (2.10) is reached in finite divergence time, and then the multi-agent system (2.1) sequentially slide along it to reach consensus asymptotically.*

Proof Consider the Lyapunov function as the following:

$$V_1 = \frac{1}{2} s^T s \tag{2.16}$$

The derivative of V_1,

$$\begin{aligned}
\dot{V}_1 &= \sum_{i=1}^{N} \dot{s}_i s_i \\
&= \sum_{i=1}^{N} s_i \left[-ksgn(s_i) + LG_2 e_{d_i} + c \sum_{j=1}^{N} a_{ij} \left(e_{d_i} - e_{d_j} \right) \right] \\
&= \sum_{i=1}^{N} \left[-k|s_i| + LG_2 e_{d_i} s_i + c s_i \sum_{j=1}^{N} a_{ij} \left(e_{d_i} - e_{d_j} \right) \right] \\
&\leqslant - \sum_{i=1}^{N} [k - (2Nc + LG_2) e_d^*] |s_i| \\
&\leqslant -\sqrt{2} [k - (2Nc + LG_2) e_d^*] V_1^{\frac{1}{2}}
\end{aligned} \tag{2.17}$$

According to the given condition $k > (2Nc + LG_2)e_d^*$, it is easy to derive that each agent can reach the sliding-mode surface $s_i = 0$ in finite time. Therefore, from (2.10) we can derive

$$\dot{x}_i = -c \sum_{j=1}^{N} a_{ij}(x_i - x_j) + d_i - \hat{d}_i \tag{2.18}$$

Combining (2.18) with the observer dynamics, yields

$$\begin{cases} \dot{x}_i &= -c \sum_{j=1}^{N} a_{ij}(x_i - x_j) + e_{d_i} \\ \dot{e}_{d_i} &= -LG_2 e_{d_i} + \dot{d}_i \\ v_i &= -c \sum_{j=1}^{N} a_{ij}(x_i - x_j) - \hat{d}_i \end{cases} \tag{2.19}$$

With the given condition that $c > 0$ and $LG_2 > 0$, it can be verified that the following system:

$$\begin{cases} \dot{x}_i &= -c \sum_{j=1}^{N} a_{ij}(x_i - x_j) + e_{d_i} \\ \dot{e}_{d_i} &= -LG_2 e_{d_i} \end{cases} \tag{2.20}$$

is exponentially stable. Since $\lim_{t \to \infty} \dot{d}_i = 0$, it can be derived that the system

$$\begin{cases} \dot{x}_i &= -c\sum_{j=1}^{N} a_{ij}(x_i - x_j) + e_{d_i} \\ \dot{e}_{d_i} &= -LG_2 e_{d_i} + \dot{d}_i \end{cases} \tag{2.21}$$

is input-to-state stable (ISS). From the system (2.21), we can derive that $\lim_{t\to\infty} e_{d_i}(t) = 0$ and $\lim_{t\to\infty}(x_i(t) - x_j(t)) = 0$, implying that the system position can reach consensus asymptotically with the designed SMC method.

Remark 2.3 From the system (2.20), we can have $\lim_{t\to\infty} e_{d_i}(t) = 0$, further $\dot{x}_i = -c\sum_{j=1}^{N} a_{ij}(x_i - x_j)$ can be obtained, according to Ren and Atkins (2007), the position $x_i(t)$ can reach consensus satisfying $\lim_{t\to\infty}\left(x_i(t) - \sum_{i=1}^{N}\alpha_i x_i(0)\right) = 0$, where $\alpha = [\alpha_1, \ldots, \alpha_n]^T$ is a non-negative left eigen vector of $-L$ associated with eigenvalue 0 with $\alpha_i \geq 0$ and $\sum_{i=1}^{N}\alpha_i = 1$. Furthermore, the velocity $v_i(t)$ can reach consensus satisfying $\lim_{t\to\infty} v_i(t) = \lim_{t\to\infty}\dot{x}(t) = 0$.

2.5 Continuous SMC via a Disturbance Observer

With the fractional-order technique, we improve the discontinuous controller above and design a new continuous SMC in this section.

The continuous protocol for MAS (2.1) is as follows:

$$u_i = -\left[c\sum_{j=1}^{N} a_{ij}(v_i - v_j) + c\sum_{j=1}^{N} a_{ij}(\hat{d}_i - \hat{d}_j) + ksig^{1/2}(s_i)\right] \tag{2.22}$$

where $sig^{1/2}(s_i) = |s_i|^{1/2}sgn(s_i)$.

Theorem 2.2 *With the fractional-order technique, the new proposed continuous controller (2.22) and disturbance observer (2.9) can guarantee that the sliding-mode surface (2.10) is reached in finite divergence time, and then the multi-agent system (2.1) sequentially slide along it to reach consensus asymptotically.*

Proof Consider the Lyapunov function as the following:

$$V_2 = \frac{1}{2}\left[sig^{1/2}(s)\right]^T sig^{1/2}(s) \tag{2.23}$$

The derivative of V_2,

$$\dot{V}_2 = \sum_{i=1}^{N}\frac{1}{2}|s_i|^{-\frac{1}{2}}\dot{s}_i sig^{1/2}(s_i)$$

$$
\begin{aligned}
&= \sum_{i=1}^{N} \frac{1}{2} sgn(s_i)\dot{s}_i \\
&= \sum_{i=1}^{N} \frac{1}{2} sgn(s_i)\left[-ksig^{1/2}(s_i) + LG_2 e_{d_i} + c\sum_{j=1}^{N} a_{ij}(e_{d_i} - e_{d_j}) \right] \\
&= -\frac{1}{2}\sum_{i=1}^{N} ksgn(s_i)sig^{1/2}(s_i) + \frac{1}{2}\sum_{i=1}^{N} sgn(s_i)\left[c\sum_{j=1}^{N} a_{ij}(e_{d_i} - e_{d_j}) + LG_2 e_{d_i} \right] \\
&\le -\frac{1}{2}k\sum_{i=1}^{N} |s_i|^{-\frac{1}{2}} sig^{1/2}(s_i) sig^{1/2}(s_i) + k(CN^2 + \frac{1}{2}LG_2)e_d^* \qquad (2.24)
\end{aligned}
$$

In light of $|s_i|^{\frac{1}{2}} = |sig^{\frac{1}{2}}(s_i)| = V_{2i}^{\frac{1}{2}}$, the inequality (2.24) can be further deduced to

$$
\begin{aligned}
\dot{V}_2 &\le -\frac{1}{2}k\sum_{i=1}^{N} V_{2i}^{-\frac{1}{2}} V_{2i} + k(CN^2 + \frac{1}{2}LG_2)e_d^* \\
&= -\frac{1}{2}k\sum_{i=1}^{N} V_{2i}^{-\frac{1}{2}} + k(CN^2 + \frac{1}{2}LG_2)e_d^* \\
&\le -\frac{1}{2}kV_2^{\frac{1}{2}} + k(CN^2 + \frac{1}{2}LG_2)e_d^* \qquad (2.25)
\end{aligned}
$$

Therefore, similar to Theorem 2.1, we can obtain that the system position and velocity will achieve consensus asymptotically with the continuous SMC method.

Remark 2.4 Throughout the three controllers, the Integral SMC is effective to remove the offset asymptotically, but it always brings some adverse effects such as chattering. To restrain the chattering phenomenon, the disturbance observer-based method helps. The discontinuous SMC via disturbance observer suppresses the chattering, but the execution of the controller is discontinuous, which is not good in practice. To balance them, the improved continuous SMC via disturbance observer is designed. In fact, the continuous SMC via disturbance observer works better for the multi-agent system (2.1).

2.6 Numerical Simulation

Numerical simulations are given to verify the effectiveness of the theoretical result in this section. There are five agents, which are denoted by $i = 1, 2, \ldots, 5$, respectively. The communication topology of the agents are given as Fig. 2.1.

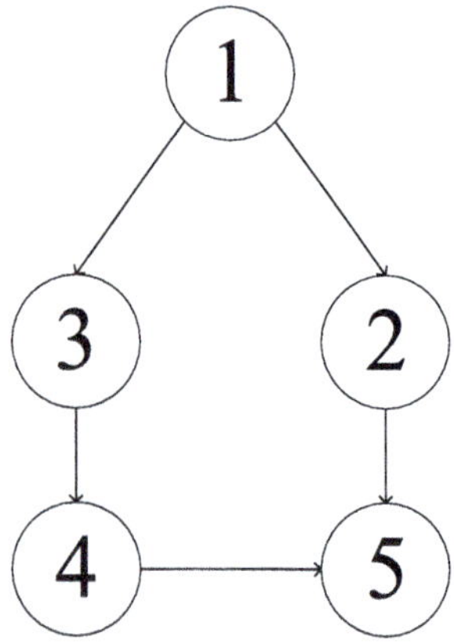

Fig. 2.1 The communication topology

In this example, we suppose there exist five agents for the multi-agent system. The dynamics of each agent is described by the MAS (2.1) and the nonlinear uncertainties $d_i(t)$ are described by $d_i(t) = e^{-t}$.

(A) Integral SMC

Consider the MAS (2.1) with the integral sliding surface (2.2) and controller (2.3): Figs. 2.2 and 2.3 describe the positions and velocities of the five agents, respectively. Figures 2.4 and 2.5 present the control inputs and sliding surfaces, respectively.

From Figs. 2.2 and 2.3, it is easy to find that the integral sliding mode controller exhibits robustness in the presence of mismatched uncertainties. However, from Fig. 2.4 it always brings adverse effects, such as overshoot and chattering.

(B) Discontinuous SMC via Disturbance Observer

To improve the control performance, the disturbance observer is introduced to alleviate the chattering phenomenon. With the proposed discontinuous sliding mode controller (2.11), the following Figs. 2.6, 2.7, 2.8 and 2.9 are derived.

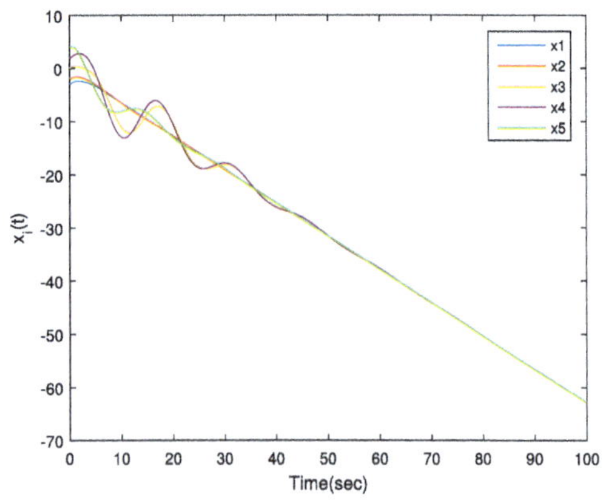

Fig. 2.2 The position x_i of each agent

Fig. 2.3 The velocity v_i of each agent

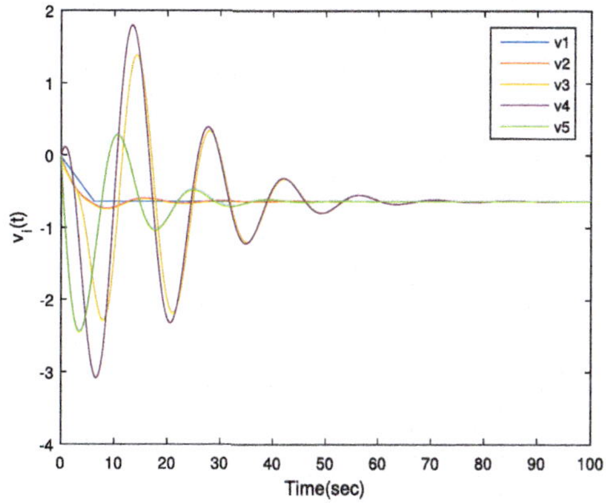

Fig. 2.4 The input u_i of each agent

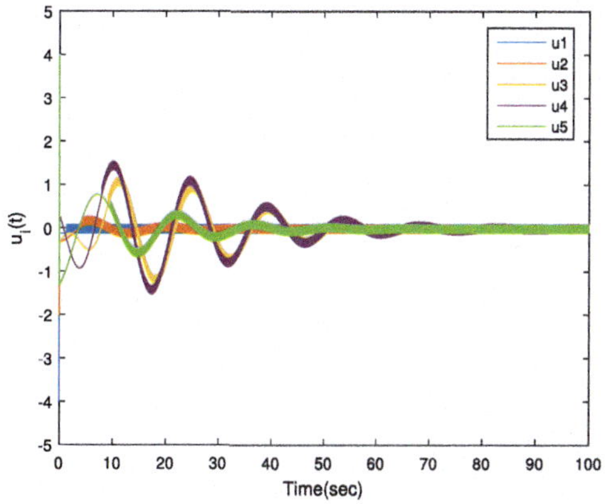

Fig. 2.5 The sliding surface s_i of each agent

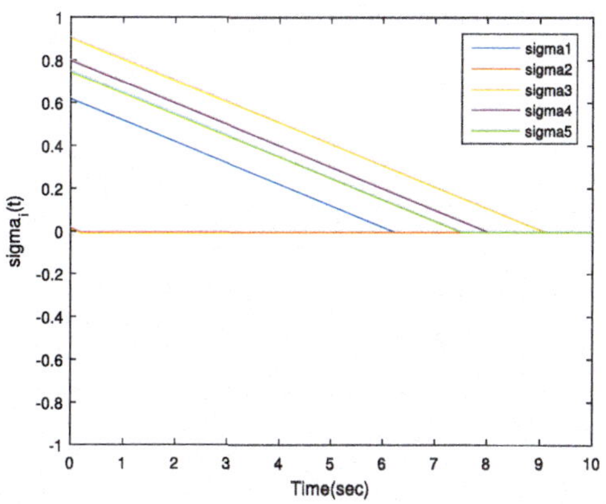

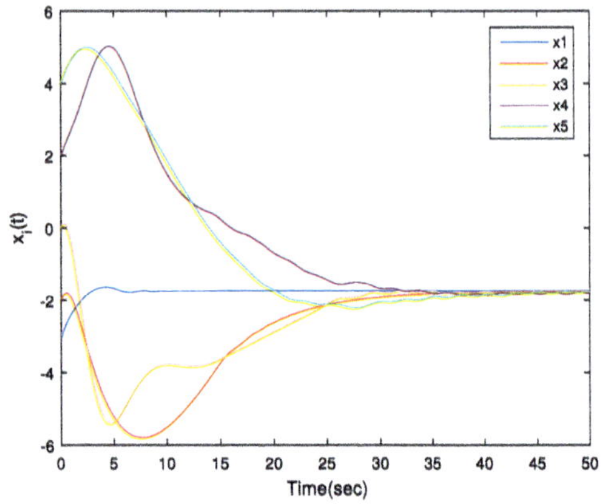

Fig. 2.6 The position x_i of each agent

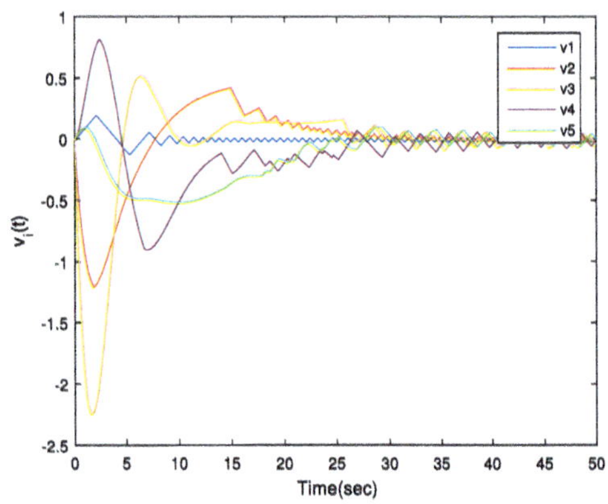

Fig. 2.7 The velocity v_i of each agent

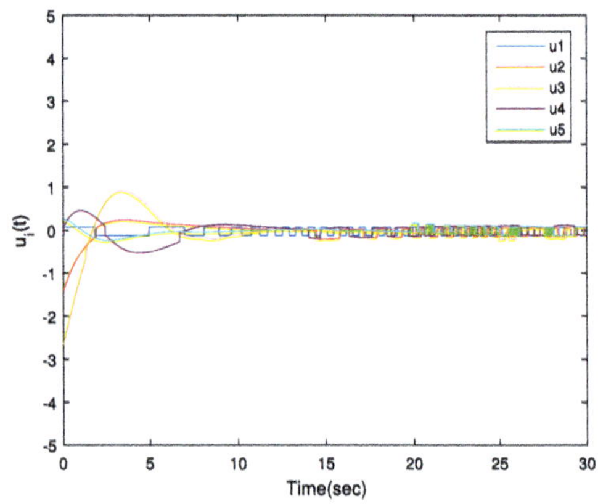

Fig. 2.8 The input u_i of each agent

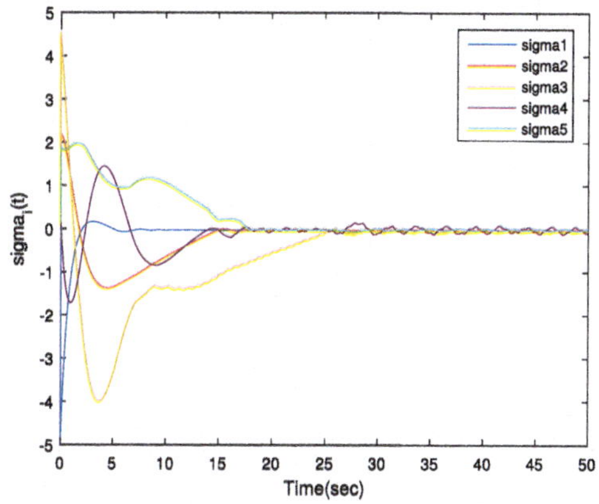

Fig. 2.9 The sliding surface s_i of each agent

Obviously, the chattering phenomenon is alleviated with the disturbance observer in comparison with the integral SMC.

(C) Continuous SMC via Disturbance Observer
To further improve the control performance, a new continuous SMC is designed with the fractional-order technique as equation (2.22) and the following Figs. 2.10, 2.11, 2.12 and 2.13 are derived.

(D) $d_i(t) = e^{-t} + v_i(t)\sin(v_i(t))$ *with continuous SMC via Disturbance Observer*
In practice, the uncertainties are mostly related with the velocities of the system. Therefore, in this section we consider the uncertainties $d_i(t) = e^{-t} + v_i(t)\sin(v_i(t))$ and derive the Figs. 2.14, 2.15, 2.16 and 2.17 with the continuous SMC via disturbance observer.

From the simulation results, it is obvious to see that the continuous SMC still works well to achieve consensus for the multi-agent system (2.1).

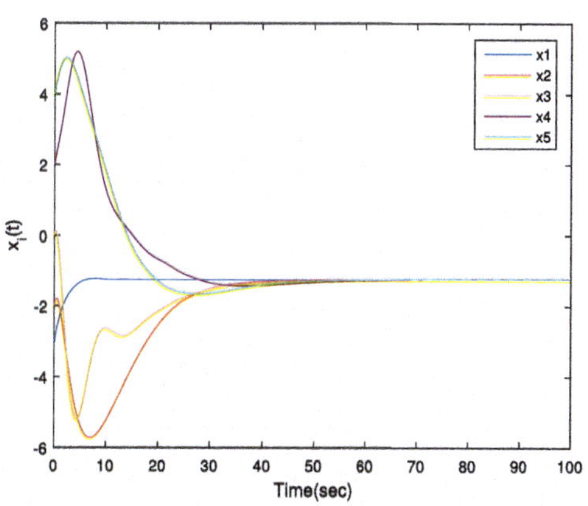

Fig. 2.10 The position x_i of each agent

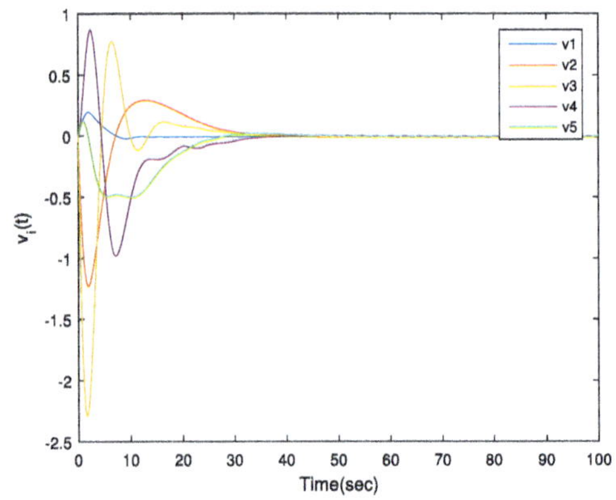

Fig. 2.11 The velocity v_i of each agent

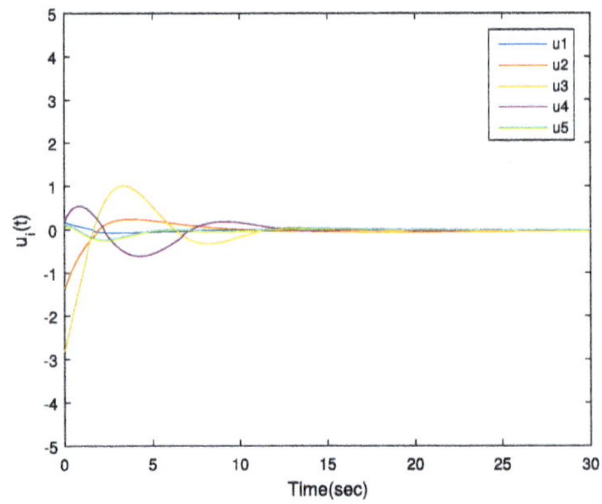

Fig. 2.12 The input u_i of each agent

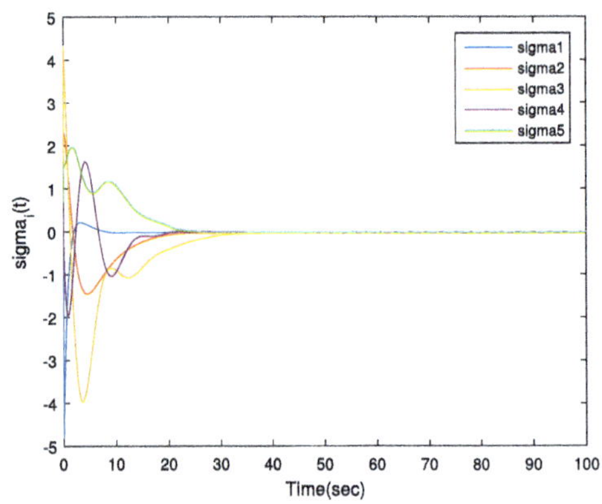

Fig. 2.13 The sliding surface s_i of each agent

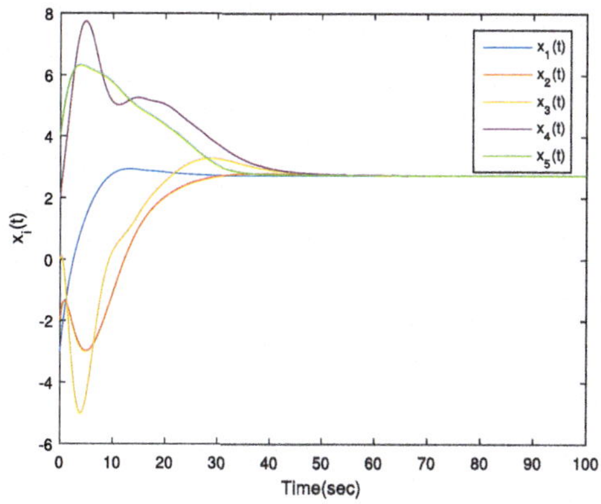

Fig. 2.14 The position x_i of each agent

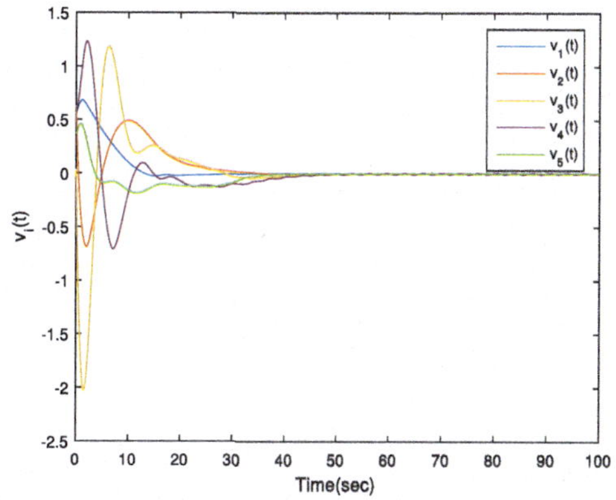

Fig. 2.15 The velocity v_i of each agent

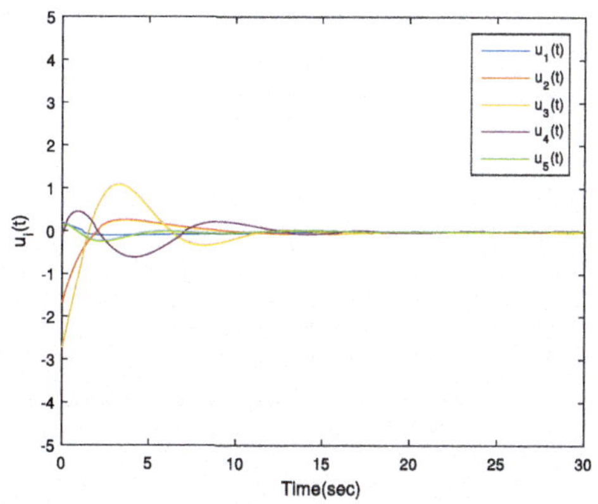

Fig. 2.16 The input u_i of each agent

2.7 Conclusion

This chapter establishes finite time sliding mode control protocols for group consensus to deal with multi-agent systems with mismatched uncertainties. In conjunction

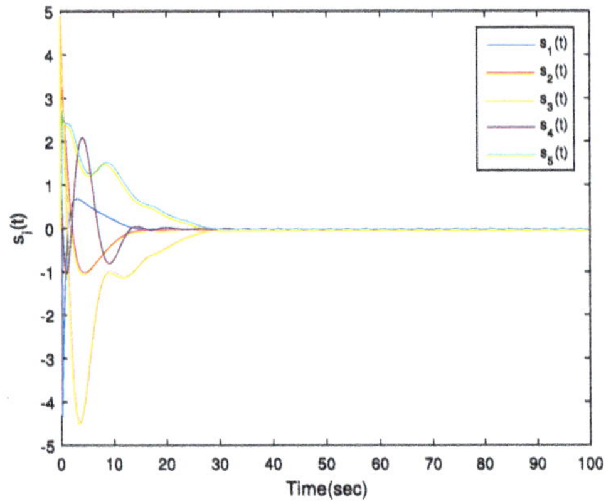

Fig. 2.17 The sliding surface s_i of each agent

with disturbance observer, the uncertainties are estimated and chattering phenomenon is alleviated. In addition, both discontinuous and continuous sliding mode controllers are proposed in this chapter. A numerical simulation is shown to verify the effectiveness of the theoretical analysis. It is inevitable to avoid time delay in practical systems. In next works, nonlinear multi-agent systems with time delay will be considered.

References

Chang, Y.-H., Chang, C.-W., Chen, C.-L., and Tao, C.-W. (2012). Fuzzy sliding-mode formation control for multirobot systems: Design and implementation, *IEEE Transactions on Systems, Man, and Cybernetics, Part B: Cybernetics***42**, 2, pp. 444–457.

Han, T., Guan, Z.-H., Liao, R.-Q., Chen, J., Chi, M., and He, D.-X. (2017). Distributed finite-time formation tracking control of multi-agent systems via ftsmc approach, *IET Control Theory and Applications***11**, 15, pp. 2585–2590.

Li, J., Yang, Y., Hua, C., and Guan, X. (2017). Fixed-time backstepping control design for high-order strict-feedback non-linear systems via terminal sliding mode, *IET Control Theory and Applications***11**, 8, pp. 1184–1193.

Lu, K., Xia, Y., Zhu, Z., and Basin, M. V. (2012). Sliding mode attitude tracking of rigid spacecraft with disturbances, *Journal of the Franklin Institute***349**, 2, pp. 413–440.

Mu, C., Xu, W., and Sun, C. (2016). On switching manifold design for terminal sliding mode control, *Journal of the Franklin Institute***353**, 7, pp. 1553–1572.

Ren, W. and Atkins, E. (2007). Distributed multi-vehicle coordinated control via local information exchange, *International Journal of Robust and Nonlinear Control***17**, 10-11, pp. 1002–1033.

Ren, C.-E. and Chen, C. L. P. (2015). Sliding mode leader-following consensus controllers for second-order non-linear multi-agent systems, *IET Control Theory and Applications***9**, 10, pp. 1544–1552.

Sam, Y. M., Osman, J. H. S., and Ghani, M. R. A. (2004). A class of proportional-integral sliding mode control with application to active suspension system, *Systems and Control Letters***51**, 3-4, pp. 217–223.

Yan, Y., Galias, Z., Yu, X., and Sun, C. (2016). Euler's discretization effect on a twisting algorithm based sliding mode control, *Automatica***68**, pp. 203–208.

Yang, J., Li, S., and Yu, X. (2013). Sliding-mode control for systems with mismatched uncertainties via a disturbance observer, *IEEE Transactions on Industrial Electronics* **60**, 1, pp. 160–169.

Yang, J., Su, J., Li, S., and Yu, X. (2014). High-order mismatched disturbance compensation for motion control systems via a continuous dynamic sliding-mode approach, *IEEE Transactions on Industrial Informatics* **10**, 1, pp. 604–614.

Yu, S. and Long, X. (2015). Finite-time consensus for second-order multi-agent systems with disturbances by integral sliding mode, *Automatica* **54**, pp. 158–165.

Zhang, J., Liu, X., Xia, Y., Zuo, Z., and Wang, Y. (2016). Disturbance observer-based integral sliding-mode control for systems with mismatched disturbances, *IEEE Transactions on Industrial Electronics* **63**, 11, pp. 7040–7048.

Zou, A.-M., Kumar, K. D., and Hou, Z.-G. (2013). Distributed consensus control for multi-agent systems using terminal sliding mode and chebyshev neural networks, *International Journal of Robust and Nonlinear Control* **23**, 3, pp. 334–357.

Chapter 3
Observer-Based Containment for a Class of Nonlinear Multi-agent Systems with Time-Delayed Protocols

Nowadays, an interesting topic is considering consensus for agents with one or multiple leaders. The case with one leader is called cooperative tracking or leader-follower consensus problem, aiming to drive the followers to approach the leader. In Zhang and Lewis (2012), a robust adaptive neural network controller was designed for each follower node such that all follower nodes ultimately synchronize to the leader node with bounded residual errors. In Hong et al. (2006), a multi-agent consensus problem with an active leader and variable interconnection topology was considered. The state of the considered leader not only kept changing but also may not be measured. To track such a leader, a neighbor-based local controller together with a neighbor-based state-estimation rule was given for each autonomous agent. The distributed tracking control problem of multiagent systems with general linear dynamics and a leader whose control input was nonzero and not available to any follower was investigated in Li et al. (2013). Based on the relative states of neighboring agents, two distributed discontinuous controllers with static and adaptive coupling gains, respectively, are designed for each follower to ensure that the states of the followers converge to the state of the leader.

In the presence of multiple leaders, the containment control problem arises, where the followers are driven into the convex hull spanned by the leaders. Containment control for multi-agent system has been widely studied. In Cao et al. (2012), distributed containment protocols were proposed for a group of mobile autonomous agents with multiple stationary or dynamic leaders for both fixed and switching topologies in directed networks. Adaptive static discontinuous and continuous controllers based on state feedback were proposed to solve the containment control problem for linear multi-agent systems in Li et al. (2015). Containment problem was transformed into asymptotic stability problem and necessary and sufficient conditions to achieve containment were given in Dong et al. (2015). Except these research for linear multi-agent systems, there also exist many fruitful results for nonlinear multi-agent systems. In Hu et al. (2016a), the containment control for multi-agent systems

Q. Wang et al., *Distributed Cooperative Control and Optimization for Multi-Agent Systems*, https://doi.org/10.1007/978-981-96-0950-5_3

with unknown nonlinear dynamics was proposed. Based on the outdated information stored in each agent's memory, a hybrid time-delayed protocol was provided to guarantee the asymptotical containment of the nonlinear multi-agent system.

Although sufficient or even necessary and sufficient conditions for multi-agent systems have been proposed to achieve containment, the state information of the agents are required available to construct the protocols. Generally, the controller based on state feedback is very powerful for both ordinary differential equations and functional differential equations if the full information of the state vectors is accessible. However, in many practical systems, only the measured output or part of state information is available. Motivated by Li et al. (2010) and Zhang et al. (2011), it is more practical to use an observer-based controller, where the measured output information is employed. In the work (Li et al. 2010), a distributed observer-type consensus protocol based on relative output measurements was proposed, which can be regarded as an extension of the traditional observer-based controller for a single system. In Hu et al. (2014), both local observer and distributed pinning observer were proposed for nonlinear multi-agent systems satisfying the Lipschitz condition, The former was constructed with the local output information and the latter was constructed with the relative output information of its neighboring agents. In Namerikawa and Yoshioka (2008), new consensus control law using observers was proposed for constrained communication, such as communication delays. In Bin and Zongli (2014), consensus problem for high-order linear multi-agent systems was solved by observer based controller with constant delays.

Although memoryless controllers have the merit that they are easy to implement, they tend to be unduly conservative. In fact, employing a feedback of the outdated information as well as the current information, the controllers provided better performance than memoryless controllers, which was referred in Moon et al. (2001). Hence, if the delay-dependent controllers are employed in containment control for multi-agent systems, an improved performance will be achieved. However, few publications are focused on containment problem of nonlinear multi-agent systems with an observer-based controller involving delays. From this point, motivated by Zhou et al. (2013), we propose the observer-based containment controller with time delay in this chapter. The main contributions are as follows: (1) The containment problem for multi-agent systems with nonlinear uncertainties is investigated; (2) The delay-dependent controllers with both current and outdated information are constructed to achieve containment; (3) Observers based on measured output information are employed rather than state information.

3.1 Graph Theory

A directed graph $G = \{V, E\}$ consists of a set of nodes $V = \{v_1, v_2, \ldots, v_N\}$ and a set of edges $E \subseteq \{(v_i, v_j), v_i, v_j \in V\}$. For an edge (v_i, v_j), node v_i is called the parent node, node v_j is the child node and v_i is a neighbor of v_j. The set of neighbors of node v_i is denoted by $N_i = \{v_j \in V : (v_j, v_i) \in E\}$. An agent is called a leader if it has no neighbors, else called a follower.

The adjacency matrix $\mathcal{A} = [a_{ij}] \in \mathbb{R}^{n\times n}$ associated with G is defined as $a_{ij} > 0$ if $(v_j, v_i) \in E$, and $a_{ij} = 0$ otherwise. In this paper, $a_{ii} = 0$. The Laplacian matrix $L = [l_{ij}] \in \mathbb{R}^{n\times n}$ associated with G is defined as $l_{ii} = \sum_{j=1, j\neq i}^{n} a_{ij}$.

3.2 Containment Controller Design

Considering a group of N identical agents with continuous-time nonlinear dynamics. Suppose that there are M followers and $N - M$ leaders. The dynamics of followers are described by

$$\begin{cases} \dot{x}_i(t) = Ax_i(t) + Bu_i(t) + Df(t, x_i(t)), \\ y_i(t) = Cx_i(t), \quad i = 1, \ldots, M \end{cases} \tag{3.1}$$

and leaders are

$$\begin{cases} \dot{x}_i(t) = Ax_i(t) + Df(t, x_i(t)), \\ y_i(t) = Cx_i(t), \quad i = M+1, \ldots, N \end{cases} \tag{3.2}$$

In this paper, the objective of containment control is to design containment protocols to drive the followers asymptotically converge to the convex hull spanned by the leaders. The problem is mathematically defined as follows:

Definition 3.1 A set $K \in \mathbb{R}^m$ is said to be convex if for any $x, y \in K$ and $\theta \in [0, 1]$, $(1-\theta)x + \theta y \in K$. The convex hull $\mathrm{Co}\{X\}$ of a finite set of points $X = \{x_1, x_2, \}$ is the minimal convex set containing all points in X. More specifically, $\mathrm{Co}\{X\} = \{\sum_{i=1}^{n} \alpha_i x_i | x_i \in X, \alpha_i \geq 0, \sum_{i=1}^{n} \alpha_i = 1\}$.

Definition 3.2 The multi-agent system described by (3.1) and (3.2) is said to achieve containment if for any given bounded initial states, there exists nonnegative constants $\alpha_j (j = M+1, \ldots, N)$ satisfying

$$\lim_{t\to\infty} (x_i(t) - \sum_{j=M+1}^{N} \alpha_j x_j(t)) = 0,$$

where $i = 1, \ldots, M$ and $\sum\limits_{j=M+1}^{N} \alpha_j = 1$.

Assumption 3.1 For each of the M followers, there exists at least one leader that has a directed path to that follower.

Assumption 3.2 For the nonlinear function $f(\cdot,\cdot)$ for each agent, it is required that, for any given nonnegative constants $\gamma_i \geq 0 (i = 1, \ldots, N-M)$ with $\sum_{i=1}^{N-M} \gamma_i = 1$, there exists a constant ρ such that for $x, y_i \in \mathbb{R}^n, i = 1, \cdots, N-M$, the inequality $\left\| f(t,x) - \sum_{i=1}^{N-M} \gamma_i f(t, y_i) \right\| \leq \rho \left\| x - \sum_{i=1}^{N-M} \gamma_i y_i \right\|$ is satisfied.

We assumed that each agent has access to the relative output measurements with respect to its neighbors, and for $i \in R, u_i = 0$. Motivated by the consensus protocols in Hu et al. (2016a), Li et al. (2010), we propose the following distributed dynamic containment controller for each follower as

$$\begin{cases} \dot{\hat{x}}_i(t) = A\hat{x}(t) + Bu_i(t) + Df(t, \hat{x}_i(t)) + \\ \qquad L \sum_{j=1}^{N} a_{ij}[(\hat{y}_i(t) - \hat{y}_j(t)) - (y_i(t) - y_j(t))], \\ u_i(t) = -\alpha K \sum_{j=1}^{N} a_{ij}(\hat{x}_i(t) - \hat{x}_j(t)) - \\ \qquad \beta K \sum_{j=1}^{N} a_{ij}(\hat{x}_i(t-\tau) - \hat{x}_j(t-\tau)) \end{cases} \tag{3.3}$$

and for leaders,

$$\dot{\hat{x}}_i(t) = A\hat{x}(t) + Df(t, \hat{x}_i(t)), i = M+1, \ldots, N \tag{3.4}$$

where $\hat{x}_i$ is the state estimate of x_i, $\hat{y}_i(t) = C\hat{x}_i(t)$ is the output of observers.

Define $x_F = [x_1^T, \ldots, x_M^T]^T, x_L = [x_{M+1}^T, \ldots, x_N^T]^T, F(x_F, t) = [f(t, x_1)^T, \ldots, f(t, x_M)^T]^T$, $F(x_L, t) = [f(t, x_{M+1})^T, \ldots, f(t, x_N)^T]^T$, $\hat{x}_F = [\hat{x}_1^T, \ldots, \hat{x}_M^T]^T$, $\hat{x}_L = [\hat{x}_{M+1}^T, \ldots, \hat{x}_N^T]^T$, $F(\hat{x}_F, t) = [f(t, \hat{x}_1)^T, \ldots, f(t, \hat{x}_M)^T]^T$, and $F(\hat{x}_L, t) = [f(t, \hat{x}_{M+1})^T, \ldots, f(t, \hat{x}_N)^T]^T$, respectively. Then, from (3.1) to (3.4) the closed-loop dynamics can be written as

$$\begin{cases} \dot{x}_F = (I_M \otimes A)x_F - \alpha(L_1 \otimes BK)\hat{x}_F - \\ \qquad \alpha(L_2 \otimes BK)\hat{x}_L - \beta(L_1 \otimes BK)\hat{x}_F(t-\tau) - \\ \qquad \beta(L_2 \otimes BK)\hat{x}_L(t-\tau) + (I_M \otimes D)F(x_F, t) \\ \dot{x}_L = (I_{N-M} \otimes A)x_L + (I_{N-M} \otimes D)F(x_L, t) \end{cases} \tag{3.5}$$

and

$$\begin{cases} \dot{\hat{x}}_F = (I_M \otimes A)\hat{x}_F - \alpha(L_1 \otimes BK)\hat{x}_F - \\ \qquad \alpha(L_2 \otimes BK)\hat{x}_L - \beta(L_1 \otimes BK)\hat{x}_F(t-\tau) - \\ \qquad \beta(L_2 \otimes BK)\hat{x}_L(t-\tau) + (I_M \otimes D)F(\hat{x}_F, t) \\ \qquad +(L_1 \otimes LC)\hat{x}_F + (L_2 \otimes LC)\hat{x}_L \\ \qquad -(L_1 \otimes LC)x_F - (L_2 \otimes LC)x_L \\ \dot{\hat{x}}_L = (I_{N-M} \otimes A)\hat{x}_L + (I_{N-M} \otimes D)F(\hat{x}_L, t) \end{cases} \tag{3.6}$$

Define variables $\zeta(t) = \hat{x}_F + (L_1^{-1}L_2 \otimes I_n)\hat{x}_L$ and $\eta(t) = x_F + (L_1^{-1}L_2 \otimes I_n)x_L$, then the above (3.5) and (3.6) can be transformed as

$$\begin{cases} \dot{\zeta}(t) = (I_M \otimes A)\zeta(t) + (I_M \otimes D)F(\hat{x}_L, \hat{x}_F, t) - \\ \qquad \alpha(L_1 \otimes BK)\zeta(t) - \beta(L_1 \otimes BK)\zeta(t-\tau) \\ \qquad +(L_1 \otimes LC)\zeta(t) - (L_1 \otimes LC)\eta(t) \\ \dot{\eta}(t) = (I_M \otimes A)\eta(t) + (I_M \otimes D)F(x_L, x_F, t) - \\ \qquad \alpha(L_1 \otimes BK)\zeta(t) - \beta(L_1 \otimes BK)\zeta(t-\tau) \end{cases} \tag{3.7}$$

where $F(\hat{x}_L, \hat{x}_F, t) = F(\hat{x}_F, t) + (L_1^{-1}L_2 \otimes I_n)F(\hat{x}_L, t)$ and $F(x_L, x_F, t) = F(x_F, t) + (L_1^{-1}L_2 \otimes I_n)F(x_L, t)$.

3.3 Containment Analysis

Theorem 3.1 *Under Assumptions 3.1 and 3.2, if the following conditions are satisfied, then it is easy to get that* $\|x_F(t) + (L_1^{-1}L_2 \otimes I_n)x_L(t)\| \to 0$ *as* $t \to \infty$, *which means the followers will converge to the convex hull spanned by multiple dynamic leaders.*

(1) Γ *is a positive diagonal matrix satisfying that* $\Gamma L_1 + L_1^T \Gamma > 0$;

(2) Giving positive constant ρ, *solve the following linear matrix inequality:*

$$\begin{bmatrix} \Xi' & PC^T & P & 0 \\ * & -2I_n & 0 & -I_n \\ * & * & -I_n & 0 \\ * & * & * & -I_n \end{bmatrix} < 0 \tag{3.8}$$

where $\Xi' = AP + PA^T - \sigma BB^T + \rho^2 DD^T + \varepsilon P$, *to obtain a positive definite matrix* P *and two positive scalars* ε *and* σ, *and set the gain matrices* $K = B^T P^{-1}$ *and* $L = -PC^T$.

(3) Substituting P, ε *and* σ *obtained from (2) into the following inequality, then solve such scalars* β *that satisfied the following linear matrix inequality:*

$$\begin{bmatrix} I_M \otimes \Xi_{11} & I_M \otimes \Xi_{12} & -\beta L_1 \otimes BB^T \\ * & I_M \otimes \Xi_{22} & -\beta L_1 \otimes BB^T \\ * & * & -\varepsilon I_M \otimes P \end{bmatrix} < 0 \tag{3.9}$$

where $\Xi_{11} = AP + PA^T - \alpha\lambda_{\min}/\gamma_{\max} BB^T - \lambda_{\min}/\gamma_{\max} PC^TCP + \rho^2 DD^T + P^TP + \varepsilon P$, $\Xi_{12} = \alpha\Gamma^{-1}L_1^T\Gamma \otimes BB^T - L_1 \otimes PC^TCP$, $\Xi_{22} = AP + PA^T + \rho^2 DD^T + P^TP$.

(4) The controller gain matrices $K = B^T P^{-1}$ *and* $L = -PC^T$, *and parameter* $\lambda_{\min}$ *is the minimum eigenvalue of matrix* $\Gamma L_1 + L_1^T\Gamma$, *and* $\gamma_{\max} = \max_{1 \le i \le M} \theta_i$.

Proof Consider the following Lyapunov function candidate:

$$\begin{aligned} V(t) = \zeta^T(t)(\Gamma \otimes P^{-1})\zeta(t) + \eta^T(t)(\Gamma \otimes P^{-1})\eta(t) \\ +\varepsilon \int_{t-\tau}^{t} \zeta^T(s)(\Gamma \otimes P^{-1})\zeta(s)ds, \end{aligned} \tag{3.10}$$

where Γ is the positive diagonal matrix $\Gamma = diag\{\gamma_1, \ldots, \gamma_M\}$ satisfies that $\Gamma L_1 + L_1^T \Gamma > 0$ and P is a symmetric positive matrix. Also, substituting the gain matrices $K = B^T P^{-1}$ and $L = -PC^T$, we can derive

$$\begin{aligned} \dot{V}(t) =& \zeta^T(t)[(\Gamma \otimes (P^{-1}A + A^T P^{-1}) \\ & - \alpha(\Gamma L_1 + L_1^T \Gamma) \otimes P^{-1}BB^T P^{-1} \\ & - (\Gamma L_1 + L_1^T \Gamma) \otimes C^T C]\zeta(t) \\ & + 2\zeta^T(t)(\Gamma \otimes P^{-1}D)F(\hat{x}_L, \hat{x}_F, t) \\ & - \zeta^T(t)[\alpha(L_1^T \Gamma \otimes P^{-1}BB^T P^{-1}) \\ & - (\Gamma L_1 \otimes C^T C)]\eta(t) \\ & - \eta^T(t)[\alpha(\Gamma L_1 \otimes P^{-1}BB^T P^{-1}) \\ & - (L_1^T \Gamma \otimes C^T C)]\zeta(t) \\ & - \beta\zeta^T(t)(\Gamma L_1 \otimes P^{-1}BB^T P^{-1})\zeta(t-\tau) \\ & - \beta\zeta^T(t-\tau)(L_1^T \Gamma \otimes P^{-1}BB^T P^{-1})\zeta(t) \\ & - \beta\eta^T(t)(\Gamma L_1 \otimes P^{-1}BB^T P^{-1})\zeta(t-\tau) \\ & - \beta\zeta^T(t-\tau)(L_1^T \Gamma \otimes P^{-1}BB^T P^{-1})\eta(t) \\ & + \eta^T(t)[\Gamma \otimes (P^{-1}A + A^T P^{-1})]\eta(t) \\ & + 2\eta^T(t)(\Gamma \otimes P^{-1}D)F(x_L, x_F, t) \\ & + \varepsilon\zeta^T(t)(\Gamma \otimes P^{-1})\zeta(t) \\ & - \varepsilon\zeta^T(t-\tau)(\Gamma \otimes P^{-1})\zeta(t-\tau) \end{aligned}$$

Based on Assumption 3.2, one has

$$\begin{cases} 2\zeta^T(t)(\Gamma \otimes P^{-1}D)F(\hat{x}_L, \hat{x}_F, t) \\ \quad \le \zeta^T(t)[\Gamma \otimes (\rho^2 P^{-1}DD^T P^{-1} + I)]\zeta(t) \\ 2\eta^T(t)(\Gamma \otimes P^{-1}D)F(x_L, x_F, t) \\ \quad \le \eta^T(t)[\Gamma \otimes (\rho^2 P^{-1}DD^T P^{-1} + I)]\eta(t) \end{cases} \tag{3.11}$$

$$\begin{aligned} \dot{V}(t) \le & \zeta^T(t)[(\Gamma \otimes (P^{-1}A + A^T P^{-1}) \\ & - \alpha(\Gamma L_1 + L_1^T \Gamma) \otimes P^{-1}BB^T P^{-1} \\ & - (\Gamma L_1 + L_1^T \Gamma) \otimes C^T C \end{aligned}$$

$$
\begin{aligned}
&+ \Gamma \otimes (\rho^2 P^{-1} D D^T P^{-1} + I)]\zeta(t) \\
&- \zeta^T(t)[\alpha(L_1^T \Gamma \otimes P^{-1} B B^T P^{-1}) \\
&- (\Gamma L_1 \otimes C^T C)]\eta(t) \\
&- \eta^T(t)[\alpha(\Gamma L_1 \otimes P^{-1} B B^T P^{-1}) \\
&- (L_1^T \Gamma \otimes C^T C)]\zeta(t) \\
&- \beta \zeta^T(t)(\Gamma L_1 \otimes P^{-1} B B^T P^{-1})\zeta(t-\tau) \\
&- \beta \zeta^T(t-\tau)(L_1^T \Gamma \otimes P^{-1} B B^T P^{-1})\zeta(t) \\
&- \beta \eta^T(t)(\Gamma L_1 \otimes P^{-1} B B^T P^{-1})\zeta(t-\tau) \\
&- \beta \zeta^T(t-\tau)(L_1^T \Gamma \otimes P^{-1} B B^T P^{-1})\eta(t) \\
&+ \eta^T(t)[\Gamma \otimes (P^{-1} A + A^T P^{-1}) \\
&+ \Gamma \otimes (\rho^2 P^{-1} D D^T P^{-1} + I)]\eta(t) \\
&+ \varepsilon \zeta^T(t)(\Gamma \otimes P^{-1})\zeta(t) \\
&- \varepsilon \zeta^T(t-\tau)(\Gamma \otimes P^{-1})\zeta(t-\tau)
\end{aligned}
$$

To simplify, define $\tilde{\zeta}(t) = (I_M \otimes P^{-1})\zeta(t)$, $\tilde{\eta}(t) = (I_M \otimes P^{-1})\eta(t)$, yields,

$$
\begin{aligned}
\dot{V}(t) \le\, & \tilde{\zeta}^T(t)[(\Gamma \otimes (AP + PA^T) - \alpha(\Gamma L_1 + L_1^T \Gamma) \otimes BB^T \\
&- (\Gamma L_1 + L_1^T \Gamma) \otimes PC^T CP \\
&+ \Gamma \otimes (\rho^2 D D^T + P^T P)]\tilde{\zeta}(t) \\
&- \zeta^T(t)[\alpha(L_1^T \Gamma \otimes BB^T) - (\Gamma L_1 \otimes PC^T CP)]\eta(t) \\
&- \eta^T(t)[\alpha(\Gamma L_1 \otimes BB^T) - (L_1^T \Gamma \otimes PC^T CP)]\zeta(t) \\
&- 2\beta \tilde{\zeta}^T(t)(\Gamma L_1 \otimes BB^T)\tilde{\zeta}(t-\tau) \\
&- 2\beta \tilde{\eta}^T(t)(\Gamma L_1 \otimes BB^T)\tilde{\zeta}(t-\tau) \\
&+ \tilde{\eta}^T(t)[\Gamma \otimes (AP + PA^T) \\
&+ \Gamma \otimes (\rho^2 D D^T + P^T P)]\tilde{\eta}(t) \\
&+ \varepsilon \tilde{\zeta}^T(t)(\Gamma \otimes P)\tilde{\zeta}(t) - \varepsilon \tilde{\zeta}^T(t-\tau)(\Gamma \otimes P)\tilde{\zeta}(t-\tau)
\end{aligned}
$$

Define $\delta(t) = [\tilde{\zeta}^T(t) \quad \tilde{\eta}^T(t) \quad \tilde{\zeta}^T(t-\tau)]^T$, yields

$$
\begin{aligned}
\dot{V}(t) \le\, & \tilde{\zeta}^T(t)[\Gamma \otimes (AP + PA^T - \alpha\lambda_{\min}/\gamma_{\max} BB^T \\
&- \lambda_{\min}/\gamma_{\max} PC^T CP + \rho^2 D D^T + P^T P + \varepsilon P)]\tilde{\zeta}(t) \\
&- 2\zeta^T(t)[\alpha(L_1^T \Gamma \otimes BB^T) - (\Gamma L_1 \otimes PC^T CP)]\eta(t) \\
&- 2\beta \tilde{\zeta}^T(t)(\Gamma L_1 \otimes BB^T)\tilde{\zeta}(t-\tau) \\
&- 2\beta \tilde{\eta}^T(t)(\Gamma L_1 \otimes BB^T)\tilde{\zeta}(t-\tau)
\end{aligned}
$$

$$
\begin{aligned}
&+ \tilde{\eta}^T(t)[\Gamma \otimes (AP + PA^T + \rho^2 DD^T + P^T P)]\tilde{\eta}(t) \\
&- \varepsilon\tilde{\zeta}^T(t-\tau)(\Gamma \otimes P)\tilde{\zeta}(t-\tau) \\
&= \delta^T(t)[I_3 \otimes (\Gamma \otimes I_n)]\Xi\delta(t) < 0,
\end{aligned}
$$

where

$$
\Xi = \begin{bmatrix} I_M \otimes \Xi_{11} & I_M \otimes \Xi_{12} & -\beta L_1 \otimes BB^T \\ * & I_M \otimes \Xi_{22} & -\beta L_1 \otimes BB^T \\ * & * & -\varepsilon I_M \otimes P \end{bmatrix}.
$$

From inequality (3.8), one can derive that $\Xi < 0$. Then, one can derive $\dot{V}(t) < 0$, which implies that $\eta(t) = x_F + (L_1^{-1}L_2 \otimes I_n)x_L \to 0$ as $t \to \infty$, then the containment is achieved. □

Remark 3.1 From inequality (3.8), using Schur Complement Lemma twice, one can get $\Xi_{11} < 0$, and substitute P, ε and σ obtained from inequality (3.8) into inequality (3.9), and then utilize Schur Complement Lemma again to get scalar β.

3.4 Numerical Simulation

In this section, we present a special case to demonstrate the effectiveness of the obtained theoretical results. The communication topologies among leaders and followers are given in Fig. 3.1, where $F_1, \ldots, F_6$ are six followers and nodes L_1 and L_2 are two leaders, denoted by numbers $1, 2, \ldots, 8$, respectively. The dynamics of the agents are given by (1), with

$$
A = \begin{bmatrix} -0.5 & 0 \\ 0.4 & -1.5 \end{bmatrix}; B = \begin{bmatrix} 1 \\ 1 \end{bmatrix};
$$

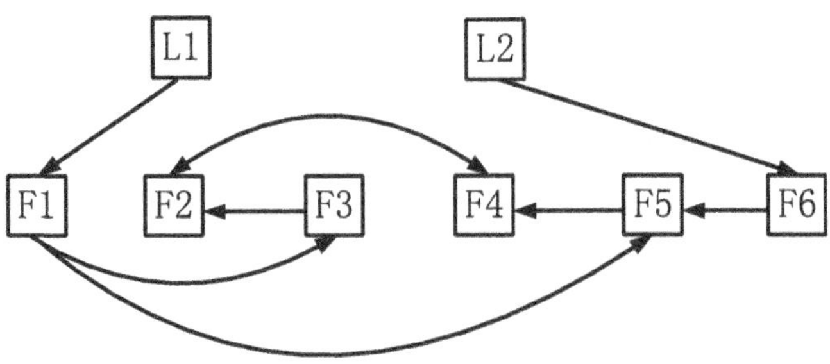

Fig. 3.1 The directed graph of agents

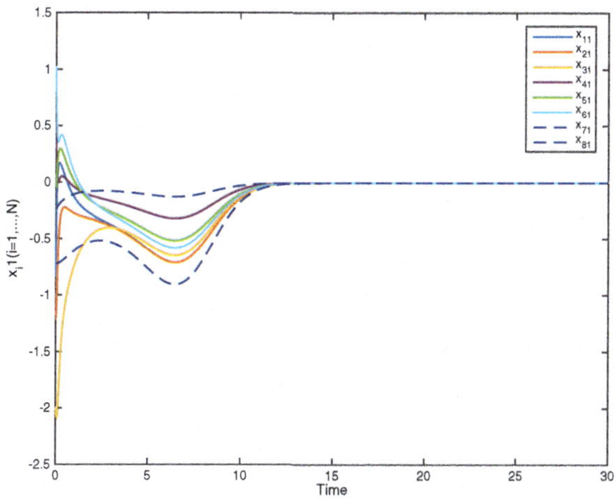

Fig. 3.2 The state trajectories of the followers

$$C = \begin{bmatrix} 1 & 0 \\ -0.2 & -1 \end{bmatrix}; D = \begin{bmatrix} 1 & -0.4 \\ 1 & -0.5 \end{bmatrix};$$

and the nonlinear uncertainties f_i for followers are given with

$$f_i = \begin{bmatrix} x_{i1}(t)\sin(0.4t) \\ x_{i2}(t)\cos(0.4t) \end{bmatrix}.$$

According to Theorem 3.1, we obtain the controller parameters as $\alpha = 3.214$, $\beta = 0.012$, $K = [1.4728\ 2.3528]$ and $L = \begin{bmatrix} -0.5078 & -0.0056 \\ -0.1071 & -0.3366 \end{bmatrix}$.

(1) The leaders nonlinear uncertainties are given with

$$f_7 = \begin{bmatrix} x_{71}\sin(0.4t) \\ x_{72}\cos(0.4t) \end{bmatrix}; f_8 = \begin{bmatrix} x_{81}\sin(0.4t) \\ x_{82}\cos(0.4t) \end{bmatrix}$$

Figures 3.2 and 3.3 represent two states of each agent, respectively. From Figs. 3.2 and 3.3, all followers and leaders can reach consensus, which is a special case of containment control.

(2) The leaders nonlinear uncertainties are given with

$$f_7 = \begin{bmatrix} \sin(t) \\ 0.5\cos(t) \end{bmatrix}; f_8 = \begin{bmatrix} \cos(t) \\ 0.5\sin(t) \end{bmatrix}$$

From Figs. 3.4 and 3.5, all followers can converge to the convex hull spanned by leaders quickly, which represents the containment is achieved.

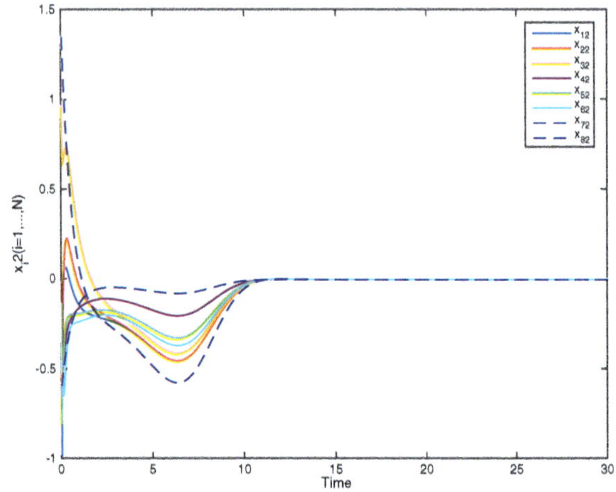

Fig. 3.3 The state trajectories of the followers

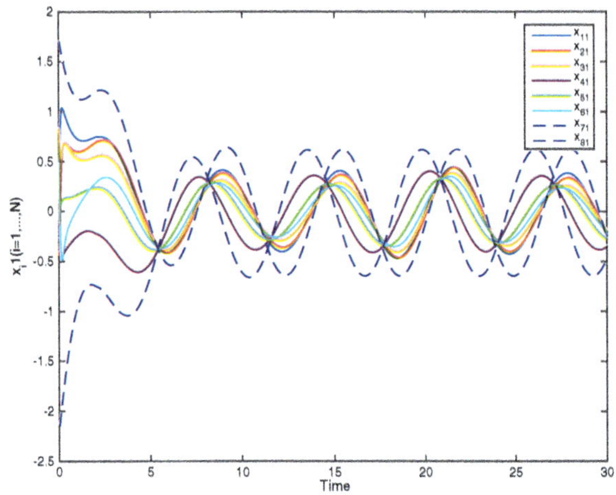

Fig. 3.4 The state trajectories of the followers

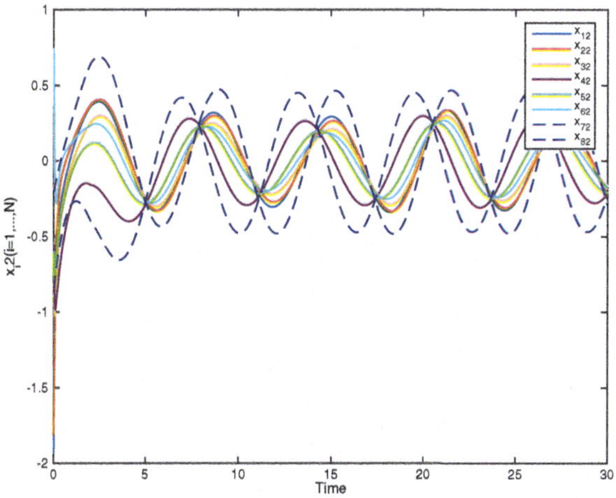

Fig. 3.5 The state trajectories of the followers

3.5 Conclusion

In this chapter, we have considered the containment control problem for nonlinear multi-agent systems. Based on observers, output information of the agents are employed to achieve containment. Therefore, a class of distributed observer-based containment protocols are proposed, and a parameter-dependent common Lyapunov function is constructed. Based on the proposed controllers, we prove that containment can be achieved for the system with nonlinear uncertainties. An interesting direction for next work is to discuss the output containment control for nonlinear multi-agent systems.

References

Bin, Z. and Zongli, L. (2014). "Consensus of high-order multi-agent systems with large input and communication delays," *Automatica***50**, 2.

Cao, Y. C., Ren, W., and Egerstedt, M. (2012). "Distributed containment control with multiple stationary or dynamic leaders in fixed and switching directed networks," *Automatica***48**, 8, pp. 1586–1597.

Dong, X., Shi, Z., Lu, G., and Zhong, Y. (2015). "Formation-containment analysis and design for high-order linear time-invariant swarm systems," *International Journal of Robust and Nonlinear Control***25**, 17, pp. 3439–3456.

Hong, Y., Hu, J., and Gao, L. (2006). "Tracking control for multi-agent consensus with an active leader and variable topology," *Automatica***42**, 7, pp. 1177–1182.

Hu, J., Cao, J., Yu, J., and Hayat, T. (2014). "Consensus of nonlinear multi-agent systems with observer-based protocols," *Systems and Control Letters***72**, pp. 71–79.

Hu, J., Cao, J., Yuan, K., and Hayat, T. (2016a). "Cooperative tracking for nonlinear multi-agent systems with hybrid time-delayed protocol," *Neurocomputing***171**, pp. 171–178.

Li, Z., Duan, Z., Chen, G., and Huang, L. (2010). "Consensus of multiagent systems and synchronization of complex networks: A unified viewpoint," *IEEE Transactions on Circuits and Systems I: Regular Papers***57**, 1, pp. 213–224.
Li, Z., Liu, X., Ren, W., and Xie, L. (2013). "Distributed tracking control for linear multiagent systems with a leader of bounded unknown input," *IEEE Transactions on Automatic Control***58**, 2, pp. 518–523.
Li, Z., Duan, Z., Ren, W., and Feng, G. (2015). "Containment control of linear multi-agent systems with multiple leaders of bounded inputs using distributed continuous controllers," *International Journal of Robust and Nonlinear Control***25**, 13, pp. 2101–2121.
Moon, Y. S., Park, P., and Kwon, W. H. (2001). "Robust stabilization of uncertain input-delayed systems using reduction method," *Automatica***37**, 2, pp. 307–312.
Namerikawa, T. and Yoshioka, C. (2008). "Consensus control of observer-based multi-agent system with communication delay," in *2008 SICE Annual Conference*, 2008 SICE Annual Conference, pp. 2414–2419.
Zhang, H. and Lewis, F. L. (2012). "Adaptive cooperative tracking control of higher-order nonlinear systems with unknown dynamics," *Automatica***48**, 7, pp. 1432–1439.
Zhang, H., Lewis, F. L., and Das, A. (2011). "Optimal design for synchronization of cooperative systems: State feedback, observer and output feedback," *IEEE Transactions on Automatic Control***56**, 8, pp. 1948–1952.
Zhou, B., Li, Z.Y. and Lin, Z. (2013). "Observer based output feedback control of linear systems with multiple input and output delays," *Automatica***7**, 49, pp. 2039–2052.

Chapter 4
Robust Output Containment Control of Multi-agent Systems with Unknown Heterogeneous Nonlinear Uncertainties in Directed Networks

In Chap. 3, state containment control is investigated for multi-agent systems. However, in some certain circumstances, state information may not be easily measurable or what we measured are just the followers' outputs. In view of these, the study about output containment is significant and meaningful.

Generally, the outputs for a system are composed of states and observable matrix. The observable matrix is always not full row rank, making the analysis and design complicated and challenging for output containment control. In Dong et al. (2013), expansion dimensions method was proposed to solve this problem for linear systems. Despite of the above, few results are concerned about output containment control for nonlinear multi-agent systems in directed networks, which motivates our research. Nonlinear uncertainties are inevitable in practical systems due to measurement error, unmodeled dynamics, external disturbances, et al. Therefore, varieties of control strategies, such as backstepping (Yao et al. 2014; Yu et al. 2014), sliding mode control (Sun et al. 2014), neural networks (Wang and Huang 2005), adaptive control (Yao et al. 2014; Yoo 2013), have been developed to deal with nonlinear systems. Robust control method (Yu et al. 2015) is also widely used to deal with nonlinear system. In this paper, robust method is selected to solve the containment problem. Compared with these control methods, robust dynamical properties can be achieved: The nominal dynamical responses can be specified by the output-feedback controller, and the influences of uncertainties can be restrained by the robust compensator. Although containment protocols for LTI multi-agent systems were proposed to achieve state containment in Dong et al. (2014) and Li et al. (2015), the relative state information was required. In our paper, only output information was required to transform the containment problem into static output feedback stabilization problem. The main contributions in the paper are as follows: Firstly, the dimensions of observable matrix is extended so that it is nonsingular. Then the containment problem is transformed into stability problem. Secondly, output feedback controller is designed to realize the stability. The nominal system constructed by the nominal controller contains desired containment properties. To restrain the external disturbances,

Q. Wang et al., *Distributed Cooperative Control and Optimization for Multi-Agent Systems*, https://doi.org/10.1007/978-981-96-0950-5_4

robust compensator method is added into the controller. A sufficient condition is given and the containment error is proved to be limited to an arbitrarily small threshold theoretically.

4.1 Mathematical Preliminaries

4.1.1 Notations

For simplicity of notation, 0 is used to denote zero matrices of appropriate size with zero vectors and zero numbers as special cases, I stands for an identity matrix with appropriate dimension, C_m^n is the combination operation in math, $x^{(n)}$ is the n-order differential operators, $\otimes$ refers to Kronecker product and $\|\cdot\|$ represents the Euclidean norm, respectively.

4.1.2 Graph Theory

A directed graph $G = \{V, E\}$ consists of a set of nodes $V = \{v_1, v_2, \dots, v_N\}$ and a set of edges $E \subseteq \{(v_i, v_j), v_i, v_j \in V\}$. For an edge (v_i, v_j), node v_i is called the parent node, node v_j is the child node and v_i is a neighbor of v_j. The set of neighbors of node v_i is denoted by $N_i = \{v_j \in V : (v_j, v_i) \in E\}$. An agent is called a leader if it has no neighbors, else called a follower.

The adjacency matrix $\mathcal{A} = [a_{ij}] \in \mathbb{R}^{n\times n}$ associated with G is defined as $a_{ij} > 0$ if $(v_j, v_i) \in E$, and $a_{ij} = 0$ otherwise. In this paper, $a_{ii} = 0$. The Laplacian matrix $L = [l_{ij}] \in \mathbb{R}^{n\times n}$ associated with G is defined as $l_{ii} = \sum_{j=1, j\neq i}^{n} a_{ij}$.

Consider a group of N agents systems, there exist M followers, labeled as agents 1 to M, and $N - M$ leaders $(N > M)$, labeled as agents $M + 1$ to N. Simply,we use $\mathcal{F} \triangleq \{1, 2, \dots, M\}$ and $\mathcal{R} \triangleq \{M + 1, M + 2, \dots, N\}$ to denote the follower set and the leader set, respectively. The Laplacian matrix corresponding to G can be partitioned as $L = \begin{bmatrix} L_1 & L_2 \\ \mathbf{0}_{(N-M)\times M} & \mathbf{0}_{(N-M)\times(N-M)} \end{bmatrix}$, where $L_1 \in \mathbb{R}^{M\times M}$ and $L_2 \in \mathbb{R}^{M\times(N-M)}$.

Assumption 4.1 For each follower, there exists at least one leader that has a directed path to the follower.

Lemma 4.1 *Under Assumption 4.1, all eigenvalues of L_1 have positive real parts, $-L_1^{-1}L_2$ is nonnegative, and each row of $-L_1^{-1}L_2$ has a sum equal to one.*

Remark 4.1 In mathematics, L_1 belongs to a class of well-known matrices, called nonsingular M-matrices. One property of such matrices is that L_1^{-1} is a nonnegative matrix. Thus, because each entry of L_2 is nonpositive, it follows that $-L_1^{-1}L_2$ is a nonnegative matrix. Note that $\begin{bmatrix} L_1 & L_2 \end{bmatrix} \begin{bmatrix} \mathbf{1}_M \\ \mathbf{1}_{N-M} \end{bmatrix} = \mathbf{0}$. Thus, it follows that $L_1 \mathbf{1}_M + L_2 \mathbf{1}_{N-M} = \mathbf{0}$, which implies that $-L_1^{-1} L_2 \mathbf{1}_{N-M} = \mathbf{1}_M$. Then we have each row sum of $-L_1^{-1} L_2$ is equal to one.

4.2 Controller Design

Suppose that the multi-agent system consists of M followers and $N - M$ leaders. Each follower with nonlinear dynamics is represented as

$$\begin{cases} \dot{x}_i(t) = Ax_i(t) + B_1 u_i(t) + B_2 \omega_i(t), \\ y_i(t) = Cx_i(t), \end{cases} \quad i \in \mathcal{F} \tag{4.1}$$

where $i = 1, \ldots, M$, $x_i(t) \in \mathbb{R}^n$ is the state of the agents, $u_i(t) \in \mathbb{R}^p$ is the control input, $y_i(t) \in \mathbb{R}^q$ is the measured output, $\omega_i(t)$ is the external disturbances.

The leaders of the concerned multi-agent system are described by

$$\begin{cases} \dot{x}_i(t) = Ax_i(t) + B_1 u_i(t), \\ y_i(t) = Cx_i(t), \end{cases} \quad i \in \mathcal{R} \tag{4.2}$$

Assumption 4.2 There are positive constants d_i and e_i such that $\|\omega_i(t)\| \le d_i + e_i \|y_i\|$, $i = 1, \ldots, M$.

Definition 4.1 A set $K \in \mathbb{R}^m$ is said to be convex if for any $x, y \in K$ and $\theta \in [0, 1]$, $(1 - \theta)x + \theta y \in K$. The convex hull $\text{Co}\{X\}$ of a finite set of points $X = \{x_1, x_2, \}$ is the minimal convex set containing all points in X. More specifically, $\text{Co}\{X\} = \{\sum_{i=1}^{n} \alpha_i x_i | x_i \in X, \alpha_i \ge 0, \sum_{i=1}^{n} \alpha_i = 1\}$.

Definition 4.2 The multi-agent system described by (4.1) and (4.2) is said to achieve output containment if for any given bounded initial states, there exists nonnegative constants $\alpha_j (j = M + 1, \ldots, N)$ satisfying

$$\lim_{t \to \infty} (y_i(t) - \sum_{j=M+1}^{N} \alpha_j y_j(t)) = 0,$$

where $\sum_{j=M+1}^{N} \alpha_j = 1$ and $i = 1, \ldots, M$.

Using Laplace transformation, from (4.1) one can have the following frequency domain transfer

$$y_i(s) = D_0^{-1}(s) N_0(s) u_i(s) + D_0^{-1}(s) L(s) \omega_i(s) \tag{4.3}$$

where $D_0^{-1}(s) N_0(s) = C(sI - A)^{-1} B_1$, $D_0^{-1}(s) L(s) = C(sI - A)^{-1} B_2$.

The robust controller is constructed as

$$u_i(s) = u_i^{nom}(s) + u_i^{rob}(s) \tag{4.4}$$

where $u_i^{nom}(s)$ stands for the nominal controller for the nominal model to obtain desired nominal performance, and $u_i^{rob}(s)$ for the robust compensator for restraining the effects of equivalent disturbance $\omega_i(s)$.

In order to restrain the effect of equivalent disturbance, the candidate robust compensator is expressed as

$$u_i^{rob*}(s) = -N_0^{-1}(s)L(s)\omega_i(s) \tag{4.5}$$

Substituting (4.4) and (4.5) into (4.3), one can further rewrite the robust compensator given by

$$u_i^{rob*}(s) = -N_0^{-1}(s)D_0(s)y_i(s) + u_i(s)$$

However, it is easy to see that this robust compensator involves high order derivative terms of $y_i(s)$, to remove the derivative terms, a robust filter, based on signal compensation technology is applied, and the robust compensator is finally given by

$$u_i^{rob}(s) = F(s)u_i^{rob*}(s) = -F(s)\left[N_0^{-1}(s)D_0(s)y_i(s) - u_i(s)\right] \tag{4.6}$$

where $F(s) = \left(\dfrac{f}{s+f}\right)^d$ is called a robust filter, with f a positive constant, and d a positive integer no less than the maximum relative degree of the elements in $N_0^{-1}(s)D_0(s)$. One can expect that by choosing sufficiently large f, it yields $u_i^{rob}(s) + \omega_i(s) \approx 0$.

Substituting (4.4) into (4.6), $u_i^{rob}(s)$ can be represented as

$$u_i^{rob}(s) = -(1 - F(s))^{-1}F(s)\left[N_0^{-1}(s)D_0(s)y_i(s) - u_i^{nom}(s)\right] \tag{4.7}$$

$N_0^{-1}(s)D_0(s)$ is known for a specific system, so the frequency domain transfer function can be realized by the state space form. Specifically, in order to realize containment, the following protocols for followers and leaders are given

$$\begin{cases} u_i(t) = u_i^{nom}(t) + u_i^{rob}(t), & i \in \mathcal{F} \\ u_i(t) = K_1 y_i(t), & i \in \mathcal{R} \end{cases} \tag{4.8}$$

where $u_i^{nom}(t) = K_1 y_i(t) + K_2 \sum_{j \in N_i} a_{ij}(y_i(t) - y_j(t))$ and $u_i^{rob}(t)$ is the state realization of (4.7), K_1 and K_2 are appropriate gain matrices to be determined later.

4.3 Controller Analysis

In this section, output containment problems for system (4.1) and (4.2) under protocol (4.8) are transformed into stability problems.

Theorem 4.1 *The multi-agent system* (4.1) *and* (4.2) *under protocol* (4.8) *achieves output containment if* A_m *are Hurwitz, where*

$$A_{mi} = \begin{bmatrix} \bar{A}_{11} + \bar{B}_{11}K_1 + \lambda_i \bar{B}_{11}K_2 & \tilde{E}_0 \\ \tilde{F}_0 + \tilde{B}_0 K_1 + \lambda_i \tilde{B}_0 K_2 & \tilde{D}_0 \end{bmatrix}.$$

Proof Since C is of full row rank, there exists $\bar{C} \in \mathbb{R}^{(n-q)\times n}$ such that $T = [C^T, \bar{C}^T]^T$ is nonsingular. Let $\bar{y}_i(t) = \bar{C}x_i(t)(i = 1, 2, \ldots, N)$,

$$TAT^{-1} = \begin{bmatrix} \bar{A}_{11} & \bar{A}_{12} \\ \bar{A}_{21} & \bar{A}_{22} \end{bmatrix}, TB_1 = \begin{bmatrix} \bar{B}_{11} \\ \bar{B}_{12} \end{bmatrix}, TB_2 = \begin{bmatrix} \bar{B}_{21} \\ \bar{B}_{22} \end{bmatrix}.$$

Then under protocol (4.8), the dynamics of multi-agent system (4.1) and (4.2) can be described as

$$\begin{cases} \dot{y}_i(t) = \bar{A}_{11}y_i(t) + \bar{A}_{12}\bar{y}_i(t) + \bar{B}_{11}K_1 y_i(t) + \bar{B}_{11}K_2 \sum\limits_{j\in N_i} a_{ij}(y_i(t) - y_j(t)) \\ \qquad + \bar{B}_{11}u_i^{rob}(t) + \bar{B}_{21}\omega_i(t), \\ \dot{\bar{y}}_i(t) = \bar{A}_{21}y_i(t) + \bar{A}_{22}\bar{y}_i(t) + \bar{B}_{12}K_1 y_i(t) + \bar{B}_{12}K_2 \sum\limits_{j\in N_i} a_{ij}(y_i(t) - y_j(t)) \\ \qquad + \bar{B}_{12}u_i^{rob}(t) + \bar{B}_{22}\omega_i(t), \end{cases}$$

in which $i \in \mathcal{F}$ and for leaders,

$$\begin{cases} \dot{y}_i(t) = \bar{A}_{11}y_i(t) + \bar{A}_{12}\bar{y}_i(t) + \bar{B}_{11}K_1 y_i(t), \\ \dot{\bar{y}}_i(t) = \bar{A}_{21}y_i(t) + \bar{A}_{22}\bar{y}_i(t) + \bar{B}_{12}K_1 y_i(t). \end{cases} \quad i \in \mathcal{R}$$

For convenience, let

$$\begin{aligned} y_F(t) &= \left[y_1^T(t), y_2^T(t), \ldots, y_M^T(t)\right]^T, \\ \bar{y}_F(t) &= \left[\bar{y}_1^T(t), \bar{y}_2^T(t), \ldots, \bar{y}_M^T(t)\right]^T, \\ y_L(t) &= \left[y_{M+1}^T(t), y_{M+2}^T(t), \ldots, y_N^T(t)\right]^T, \\ \bar{y}_L(t) &= \left[\bar{y}_{M+1}^T(t), \bar{y}_{M+2}^T(t), \ldots, \bar{y}_N^T(t)\right]^T, \\ u_F^{rob}(t) &= \left[(u_1^{rob}(t))^T, (u_2^{rob}(t))^T, \ldots, (u_M^{rob}(t))^T\right]^T, \\ \Omega(t) &= \left[\omega_1^T(t), \omega_2^T(t), \ldots, \omega_M^T(t)\right]^T, \end{aligned}$$

Further more, the system can be transformed into

$$\begin{cases} \dot{y}_F(t) = (I_M \otimes (\bar{A}_{11} + \bar{B}_{11}K_1) + L_1 \otimes \bar{B}_{11}K_2)y_F(t) + (I_M \otimes \bar{A}_{12})\bar{y}_F(t) \\ \qquad +(L_2 \otimes \bar{B}_{11}K_2)y_L(t) + (I_M \otimes \bar{B}_{11})u_F^{rob}(t) + (I_M \otimes \bar{B}_{21})\Omega(t), \\ \dot{\bar{y}}_F(t) = (I_M \otimes (\bar{A}_{21} + \bar{B}_{12}K_1) + L_1 \otimes \bar{B}_{12}K_2)y_F(t) + (I_M \otimes \bar{A}_{22})\bar{y}_F(t) \\ \qquad +(L_2 \otimes \bar{B}_{12}K_2)y_L(t) + (I_M \otimes \bar{B}_{12})u_F^{rob}(t) + (I_M \otimes \bar{B}_{22})\Omega(t), \end{cases} \tag{4.9}$$

$$\begin{cases} \dot{y}_L(t) = (I_{N-M} \otimes (\bar{A}_{11} + \bar{B}_{11}K_1) + L_1 \otimes \bar{B}_{11}K_2)y_L(t) + (I_{N-M} \otimes \bar{A}_{12})\bar{y}_L(t), \\ \dot{\bar{y}}_L(t) = (I_{N-M} \otimes (\bar{A}_{21} + \bar{B}_{12}K_1) + L_1 \otimes \bar{B}_{12}K_2)y_L(t) + (I_{N-M} \otimes \bar{A}_{22})\bar{y}_L(t), \end{cases} \tag{4.10}$$

For $i \in \mathcal{F}$, define the output error $\psi_i(t) = \sum_{j \in N_i} a_{ij}(y_i(t) - y_j(t))$ and $\Psi(t) = [\psi_1^T(t), \psi_2^T(t), \ldots, \psi_M^T(t)]^T$.

It is easy to get

$$\Psi(t) = (L_1 \otimes I)y_F(t) + (L_2 \otimes I)y_L(t) \tag{4.11}$$

According to Definition 4.2, output containment means $\lim_{t\to\infty}\Psi(t) = 0$, implying

$$\lim_{t\to\infty}(y_F(t) + (L_1^{-1}L_2 \otimes I)y_L(t)) = 0$$

From (4.9) and (4.10), one knows that $y_i(t)$ is coupled with $\bar{y}_i(t)$. It is noticed that $y_i(t)$ is only related with the observable composition of $(\bar{A}_{22}, \bar{A}_{12})$, so if $(\bar{A}_{22}, \bar{A}_{12})$ is not completely observable, then there exists a nonsingular matrix $\tilde{T}$ such that

$$(\tilde{T}^{-1}\bar{A}_{22}\tilde{T}, \bar{A}_{12}\tilde{T}) = \left(\begin{bmatrix} \tilde{D}_0 & 0 \\ \tilde{D}_1 & \tilde{D}_2 \end{bmatrix}, \begin{bmatrix} \tilde{E}_0 & 0 \end{bmatrix}\right),$$

$$\tilde{T}^{-1}\bar{A}_{21} = [\tilde{F}_0^T, \tilde{F}_1^T]^T, \qquad \tilde{T}^{-1}\bar{B}_{12} = [\tilde{B}_0^T, \tilde{B}_1^T]^T, \qquad \tilde{T}^{-1}\bar{B}_{22} = [\tilde{B}_2^T, \tilde{B}_3^T]^T,$$

where $(\tilde{D}_0, \tilde{E}_0)$ is completely observable. Define $\tilde{y}_i(t) = \tilde{T}^{-1}\bar{y}_i(t) = [\tilde{y}_{io}(t), \tilde{y}_{i\bar{o}}(t)]^T$ $(i = 1, 2, \ldots, N)$,

$$\tilde{y}_{Fo}(t) = \left[\tilde{y}_{1o}^T(t), \tilde{y}_{2o}^T(t), \ldots, \tilde{y}_{Mo}^T(t)\right]^T,$$
$$\tilde{y}_{F\bar{o}}(t) = \left[\tilde{y}_{1\bar{o}}^T(t), \tilde{y}_{2\bar{o}}^T(t), \ldots, \tilde{y}_{M\bar{o}}^T(t)\right]^T,$$
$$\tilde{y}_{Lo}(t) = \left[\tilde{y}_{M+1o}^T(t), \tilde{y}_{M+2o}^T(t), \ldots, \tilde{y}_{No}^T(t)\right]^T,$$
$$\tilde{y}_{L\bar{o}}(t) = \left[\tilde{y}_{M+1\bar{o}}^T(t), \tilde{y}_{M+2\bar{o}}^T(t), \ldots, \tilde{y}_{N\bar{o}}^T(t)\right]^T,$$

Then the multi-agent system can be further transformed into

$$\begin{cases} \dot{y}_F(t) &= (I_M \otimes (\bar{A}_{11} + \bar{B}_{11}K_1) + L_1 \otimes \bar{B}_{11}K_2)y_F(t) + (I_M \otimes \tilde{E}_0)\tilde{y}_{Fo}(t) \\ &\quad +(L_2 \otimes \bar{B}_{11}K_2)y_L(t) + (I_M \otimes \bar{B}_{11})u_F^{rob}(t) + (I_M \otimes \bar{B}_{21})\Omega(t), \\ \dot{\tilde{y}}_{Fo}(t) &= (I_M \otimes (\tilde{F}_0 + \tilde{B}_0K_1) + L_1 \otimes \tilde{B}_0K_2)y_F(t) + (I_M \otimes \tilde{D}_0)\tilde{y}_{Fo}(t)) \\ &\quad +(L_2 \otimes \tilde{B}_0K_2)y_L(t) + (I_M \otimes \tilde{B}_0)u_F^{rob}(t) + (I_M \otimes \tilde{B}_2)\Omega(t), \\ \dot{\tilde{y}}_{F\bar{o}}(t) &= (I_M \otimes (\tilde{F}_1 + \tilde{B}_1K_1) + L_1 \otimes \tilde{B}_1K_2)y_F(t) + (I_M \otimes \tilde{D}_1)\tilde{y}_{Fo}(t) \\ &\quad +(I_M \otimes \tilde{D}_2)\tilde{y}_{F\bar{o}}(t) + (L_2 \otimes \tilde{B}_1K_2)y_L(t) + (I_M \otimes \tilde{B}_1)u_F^{rob}(t) \\ &\quad +(I_M \otimes \tilde{B}_3)\Omega(t), \end{cases} \tag{4.12}$$

$$\begin{cases} \dot{y}_L(t) &= (I_{N-M} \otimes (\bar{A}_{11} + \bar{B}_{11}K_1))y_F(t) + (I_{N-M} \otimes \tilde{E}_0)\tilde{y}_{Lo}(t), \\ \dot{\tilde{y}}_{Lo}(t) &= (I_{N-M} \otimes (\tilde{F}_0 + \tilde{B}_0K_1))y_L(t) + (I_{N-M} \otimes \tilde{D}_0)\tilde{y}_{Lo}(t)), \\ \dot{\tilde{y}}_{L\bar{o}}(t) &= (I_{N-M} \otimes (\tilde{F}_1 + \tilde{B}_1K_1))y_L(t) + (I_{N-M} \otimes \tilde{D}_1)\tilde{y}_{Lo}(t) \\ &\quad +(I_{N-M} \otimes \tilde{D}_2)\tilde{y}_{L\bar{o}}(t), \end{cases} \tag{4.13}$$

For $i \in \mathcal{F}$, define $\phi_i(t) = \sum_{j \in N_i} a_{ij}(\tilde{y}_{io}(t) - \tilde{y}_{jo}(t))$, and $\Phi(t) = \left[\phi_1^T(t), \phi_2^T(t), \ldots, \phi_M^T(t)\right]^T$.

Then, we can get

$$\Phi(t) = (L_1 \otimes I)\tilde{y}_{Fo}(t) + (L_2 \otimes I)\tilde{y}_{Lo}(t), \tag{4.14}$$

Taking the derivative of (4.11) and (4.14) with respect to t respectively, one can obtain

$$\begin{cases} \dot{\Psi}(t) = (L_1 \otimes I)\dot{y}_F(t) + (L_2 \otimes I)\dot{y}_L(t), \\ \dot{\Phi}(t) = (L_1 \otimes I)\dot{\tilde{y}}_{Fo}(t) + (L_2 \otimes I)\dot{\tilde{y}}_{Lo}(t), \end{cases} \tag{4.15}$$

Then substituting (4.12) and (4.13) into the above equation (4.15), one can get

$$\begin{cases} \dot{\Psi}(t) = (I_M \otimes (\bar{A}_{11} + \bar{B}_{11}K_1) + L_1 \otimes \bar{B}_{11}K_2)\Psi(t) + (I_M \otimes \tilde{E}_0)\Phi(t) \\ \qquad +(I_M \otimes \bar{B}_{11})(L_1 \otimes I)u_F^{rob}(t) + (I_M \otimes \bar{B}_{21})(L_1 \otimes I)\Omega(t), \\ \dot{\Phi}(t) = (I_M \otimes (\tilde{F}_0 + \tilde{B}_0K_1) + L_1 \otimes \tilde{B}_0K_2)\Psi(t) + (I_M \otimes \tilde{D}_0)\Phi(t) \\ \qquad +(I_M \otimes \tilde{B}_0)(L_1 \otimes I)u_F^{rob}(t) + (I_M \otimes \tilde{B}_2)(L_1 \otimes I)\Omega(t). \end{cases} \tag{4.16}$$

The nominal system is as the following

$$\begin{cases} \dot{\Psi}_n(t) = (I_M \otimes (\bar{A}_{11} + \bar{B}_{11}K_1) + L_1 \otimes \bar{B}_{11}K_2)\Psi_n(t) + (I_M \otimes \tilde{E}_0)\Phi_n(t), \\ \dot{\Phi}_n(t) = (I_M \otimes (\tilde{F}_0 + \tilde{B}_0K_1) + L_1 \otimes \tilde{B}_0K_2)\Psi_n(t) + (I_M \otimes \tilde{D}_0)\Phi_n(t). \end{cases} \tag{4.17}$$

Obviously, if $A_{mi}(i=1,\ldots,M)$ given in Theorem 4.1 is Hurwitz, then $\lim_{t\to\infty}\Psi_n(t)=0$, which means the nominal system achieves output containment. Next, we prove the error between original system and nominal system tends to zero.

To simplify, define $g(t)=\begin{bmatrix}\Psi(t)\\ \Phi(t)\end{bmatrix}$, $g_n(t)=\begin{bmatrix}\Psi_n(t)\\ \Phi_n(t)\end{bmatrix}$, and $g_e(t)=g(t)-g_n(t)$,

$$A_m=\begin{bmatrix} I_M\otimes(\bar{A}_{11}+\bar{B}_{11}K_1)+L_1\otimes\bar{B}_{11}K_2 & I_M\otimes\tilde{E}_0\\ I_M\otimes(\tilde{F}_0+\tilde{B}_0K_1)+L_1\otimes\tilde{B}_0K_2 & I_M\otimes\tilde{D}_0\end{bmatrix},$$

$$B_\omega=\begin{bmatrix}(I_M\otimes\bar{B}_{21})(L_1\otimes I)\\ (I_M\otimes\tilde{B}_2)(L_1\otimes I)\end{bmatrix},$$

yields, one can get

$$\dot{g}_e(t)=A_m g_e(t)+B_\omega[1-F(s)]\Omega(t).$$

Introducing $q(t)$ and $z(t)$

$$q(t)=\frac{1}{s+h}[1-F(s)]\Omega(t), \tag{4.18}$$

then

$$\begin{cases} g_e(t)=z(t)+B_\omega q(t),\\ \dot{z}(t)=A_m z(t)+(A_m+hI)B_\omega(t)q(t)\end{cases} \tag{4.19}$$

The following is to prove $g_e(t)$ converge to 0. By Assumption 4.2,

$$\begin{aligned}\|\Omega(t)\| &\le \sqrt{M}\max_{i=1,\ldots,M}\|\delta_i\|\le\sqrt{M}\max_{i=1,\ldots,M}(d_i+e_i\|y_i(t)\|)\\
&\le\sqrt{M}\max_{i=1,\ldots,M}(d_i+e_i\|y_F(t)\|)\\
&\le\sqrt{M}\max_{i=1,\ldots,M}\left(d_i+e_i\|(L_1^{-1}\otimes I)\Psi(t)-(L_1^{-1}L_2\otimes I)y_L\|\right)\\
&\le\sqrt{M}\max_{i=1,\ldots,M}\left(d_i+e_i\|(L_1^{-1}\otimes I)\Psi(t)\|+e_i\|(L_1^{-1}L_2\otimes I)y_L\|\right)\\
&\le\alpha\|\Psi(t)\|+\beta\\
&\le\alpha\|g_e(t)\|+\alpha\|\Psi_n(t)\|+\beta\end{aligned} \tag{4.20}$$

where $\alpha\ge\sqrt{M}\max_{i=1,\ldots,M}(e_i\|(L_1^{-1}\otimes I)\|)$ and $\beta\ge\sqrt{M}\max_{i=1,\ldots,M}d_i+\sqrt{M}\|C\|\max_{i=M+1,\ldots,N}r_i$. Thus, substituting (4.18) into (4.20), we get

$$\|q(t)\| \leq \|\frac{1}{s+h}[1-F(s)]\|(\alpha\|g_e(t)\|+\alpha\|\Psi_n(t)\|+\beta) \tag{4.21}$$

Also, we know that

$$\left\|\frac{1}{s+h}[1-F(s)]\right\| \leq \left\|\frac{1}{s+h}(1-\frac{f^d}{(s+f)^d})\right\| \leq \left\|\frac{ds}{s+h}\frac{(s+f)^{d-1}}{(s+f)^d}\right\| \leq \left\|\frac{d}{f}\right\| \tag{4.22}$$

and

$$\|\alpha\|g_e(t)\|+\alpha\|\Psi_n(t)\|+\beta\| \leq \alpha\|z(t)\|+\alpha\|B_w\|\|q(t)\|+\alpha\|\Psi_n(t)\|+\beta \tag{4.23}$$

Choosing $c = d\max\{\alpha, \alpha B_w\}$ and $\mu = \|\Psi_n(t)\| + \left\|\frac{\beta}{\alpha}\right\|$. Combining (4.21), (4.22) and (4.23) we can easily get

$$\begin{aligned}\|q(t)\| &\leq \left\|\frac{1}{s+h}[1-F(s)]\right\|(\alpha\|g_e(t)\|+\alpha\|\Psi_n(t)\|+\beta)\\ &\leq \frac{c}{f}(\|z(t)\|+\|q(t)\|+\mu)\end{aligned}$$

Furthermore, if $f \gg c$, we can get

$$\|q(t)\| \leq \frac{c}{f-c}(\|z(t)\|+\mu) \tag{4.24}$$

Choosing the Lyapunov function candidate

$$V = z^T P z$$

where P is symmetric positive definite constant matrices, and P satisfies

$$PA_m + A_m^T P = -Q, \quad Q = Q^T > 0$$

Then we can get the derivative of V as the following

$$\begin{aligned}\dot{V} &= \dot{z}^T P z + z^T P \dot{z}\\ &= -z^T Q z + 2z^T P(A_m + hI)B_\omega q(t)\\ &\leq -z^T Q z + \zeta_z\|z(t)\|\|q(t)\|\\ &\leq -\lambda_q\|z(t)\|^2 + \zeta_z\|z(t)\|\|q(t)\|\end{aligned} \tag{4.25}$$

where $\zeta_z = \|2P(A_m + hI)B_\omega\|$ and $\lambda_q = \min_i \lambda_i(Q)$

Substituting (4.24) into (4.25), we get

$$\dot{V} \leq -\lambda_q\|z(t)\|^2 + \zeta_z\|z(t)\|\frac{c}{f-c}(\|z(t)\|+\mu)$$

$$\leq -\lambda_q \|z(t)\| \left[\|z(t)\| - \frac{\zeta_z}{\lambda_q} \frac{c}{f - c} (\|z(t)\| + \mu) \right]$$

If f is sufficiently large to make $\|z\| \geq \dfrac{2c\mu\zeta_z}{f\lambda_q - 2c\zeta_z}$, then it is easy to get $\dot{V} \leq 0$.

Then, it is easily to know that both $q(t)$ and $z(t)$ converge to 0, from (4.19), yields $g_e(t)$ converge to 0, which means $g(t)$ converge to $g_n(t)$. The system $g_n(t)$ is nominal, similar to Dong et al. (2015), according to Theorem 1, the output containment is achieved.

4.4 Numerical Simulation

In this section, numerical simulations are performed to show the effectiveness of the proposed control algorithm. Considering the containment control for eight agents with two leaders and six followers(see Fig. 4.1), marking the followers with numbers $1, 2, \ldots, 6$ and $7, 8$ for two leaders,respectively.

The dynamics of each agent by Eqs. (4.1) and (4.2) with

$$x_i = \begin{bmatrix} x_{i1} \\ x_{i2} \\ x_{i3} \\ x_{i4} \end{bmatrix}, A = \begin{bmatrix} -1 & 0 & 0 & 0 \\ -1 & -2 & 0 & 0 \\ -1 & 0 & -2 & 0 \\ 0 & -1 & 1 & -2 \end{bmatrix}, B_1 = \begin{bmatrix} 1 \\ 0 \\ 3 \\ 2 \end{bmatrix},$$

$$B_2 = \begin{bmatrix} 1\,0\,0\,0 \\ 0\,1\,0\,0 \\ 0\,0\,1\,0 \\ 0\,0\,0\,1 \end{bmatrix}, C = \begin{bmatrix} 1\ 1\ 0\ 0 \end{bmatrix}.$$

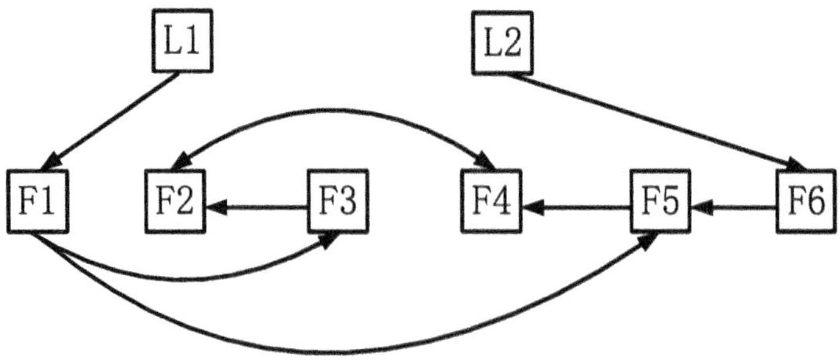

Fig. 4.1 The directed graph that characterizes the interaction among the two leaders and six followers, $L1$, $L2$ denotes two leaders, and $F1, \ldots, F6$, denotes the followers

For followers, the uncertainties are given respectively:

$$f_1 = \begin{bmatrix} 0 \\ 0.1sin(t) \\ 0 \\ 0.1sin(x_{13}) \end{bmatrix}, f_2 = \begin{bmatrix} 0 \\ 0.2cos(t) \\ 0 \\ 0.2cos(x_{23}) \end{bmatrix}, f_3 = \begin{bmatrix} 0 \\ 0 \\ 0 \\ 0.3sin(x_{33}) \end{bmatrix},$$

$$f_4 = \begin{bmatrix} 0 \\ 0.4cos(t) \\ 0 \\ 0.4cos(x_{43}) \end{bmatrix}, f_5 = \begin{bmatrix} 0 \\ 0 \\ 0 \\ 0.5sin(x_{53}) \end{bmatrix}, f_6 = \begin{bmatrix} 0 \\ 0 \\ 0 \\ 0.5sin(x_{63}) \end{bmatrix}.$$

and the leaders control inputs are designed as $u_7 = K_1 y_7 + 2sin(t)$ and $u_8 = K_1 y_8 + cos(t)$.

To illustrate Theorem 4.1, select $K_1 = -3.6613$, $K_2 = -10.8215$, $f = 100, d = 1$, and according to the proposed controller, the output trajectories $y_i(t)$ of the agents are drawn in Fig. 4.6, respectively.

From the simulation results above, the same containment control protocol is applied to multi-agent systems with different types of uncertainties. It is obvious that all followers can converge to the convex hull spanned by leaders quickly which represents the proposed containment control protocol is robust and effective.

4.5 Conclusion

In this chapter, we have considered the output containment control problem for multi-agent systems with uncertainties and multiple leaders whose control inputs are nonzero and time-varying. Through the extensive dimensions of observable matrix, the containment problem is transformed into stability problem of subsystems. Then output feedback controller with robust compensator is designed to realize the stability and restrain the external disturbances. The next work is to consider the case there exists limitation in communication and computation.

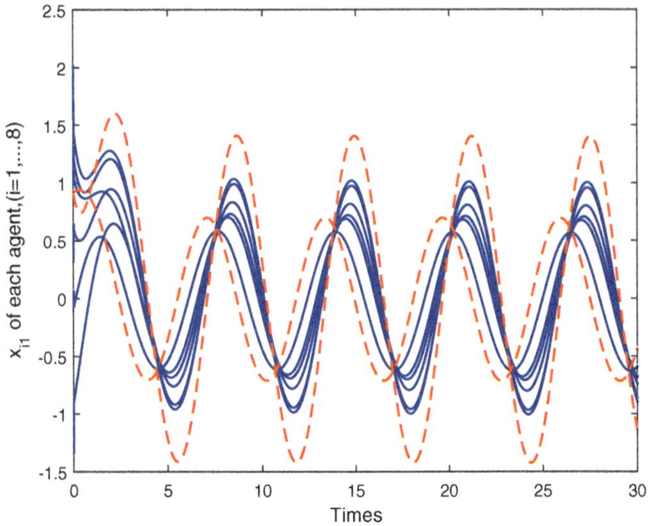

Fig. 4.2 The state x_{i1} of each agent (The two dashed lines represent the leaders and the six solid lines represent the followers)

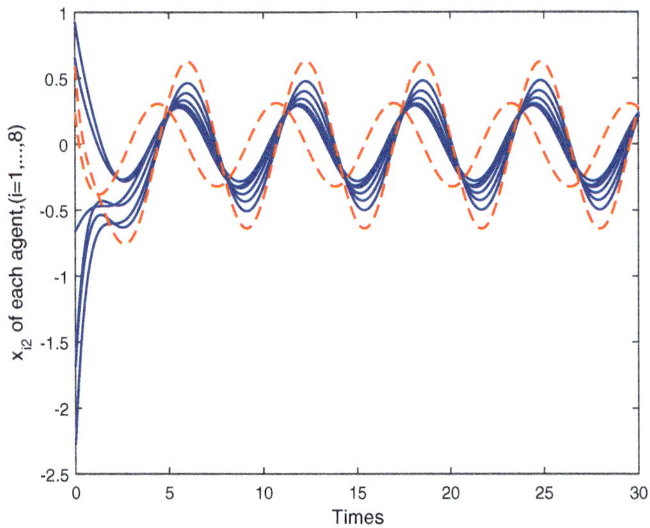

Fig. 4.3 The state x_{i2} of each agent (The two dashed lines represent the leaders and the six solid lines represent the followers)

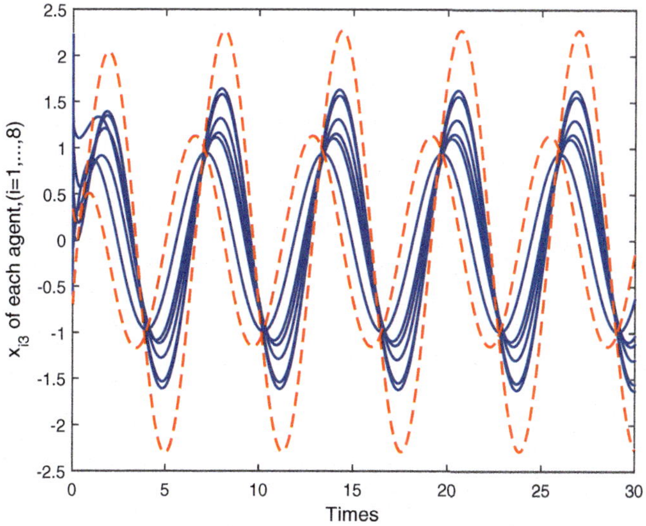

Fig. 4.4 The state x_{i3} of each agent (The two dashed lines represent the leaders and the six solid lines represent the followers)

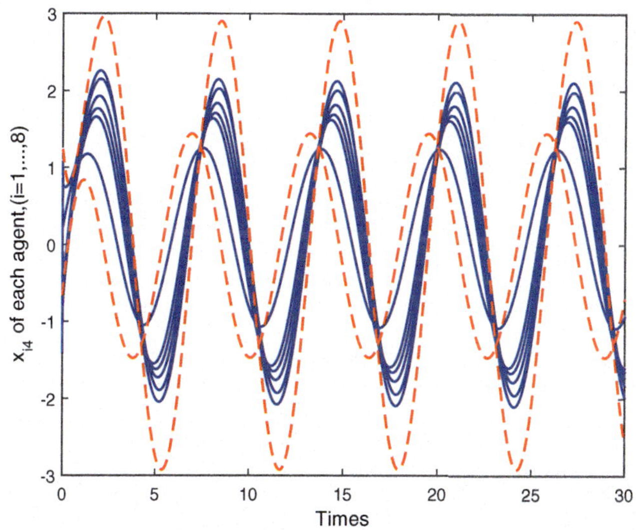

Fig. 4.5 The state x_{i4} of each agent (The two dashed lines represent the leaders and the six solid lines represent the followers)

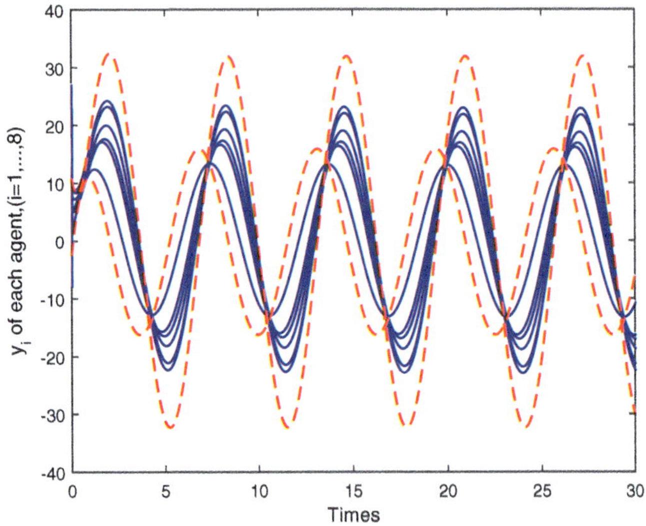

Fig. 4.6 The output y_i of each agent (The two dashed lines represent the leaders and the six solid lines represent the followers)

References

Dong, X., Shi, Z., Lu, G., & Zhong, Y. (2013). Output Containment Control for High-Order Linear Time-Invariant Swarm Systems. In IEEE International Conference on Systems, Man, and Cybernetics (pp.285–290). Manchester, UK.

Dong, X. W., Xi, J. X., Lu, G., and Zhong, Y. S. (2014). Containment analysis and design for high-order linear time-invariant singular swarm systems with time delays, *International Journal of Robust and Nonlinear Control* **24**, 7, pp. 1189–1204.

Dong, X., Shi, Z., Lu, G., and Zhong, Y. (2015). Formation-containment analysis and design for high-order linear time-invariant swarm systems, *International Journal of Robust and Nonlinear Control* **25**, 17, pp. 3439–3456.

Li, Z., Duan, Z., Ren, W., and Feng, G. (2015). Containment control of linear multi-agent systems with multiple leaders of bounded inputs using distributed continuous controllers, *International Journal of Robust and Nonlinear Control* **25**, 13, pp. 2101–2121.

Sun, C., Mu, C., and Zhang, R. (2014). *Hypersonic vehicles terminal sliding mode control technology* (Chinese Science Press, Beijing).

Wang, D. and Huang, J. (2005). Neural network-based adaptive dynamic surface control for a class of uncertain nonlinear systems in strict-feedback form, *IEEE Transactions on Neural Networks* **1**, 16, pp. 195–202.

Yao, J. Y., Jiao, Z. X., and Ma, D. W. (2014a). Extended-state-observer-based output feedback nonlinear robust control of hydraulic systems with backstepping, *IEEE Transactions on Industrial Electronics* **61**, 11, pp. 6285–6293.

Yao, J. Y., Jiao, Z. X., Ma, D. W., and Yan, L. (2014b). High-accuracy tracking control of hydraulic rotary actuators with modeling uncertainties, *IEEE-ASME Transactions on Mechatronics* **19**, 2, pp. 633–641.

Yoo, S. J. (2013). Distributed adaptive containment control of uncertain nonlinear multi-agent systems in strict-feedback form, *Automatica* **49**, 7, pp. 2145–2153.

Yu, Y., Sun, C. Y., and Jiao, Z. (2014). Robust-decentralized tracking control for a class of uncertain mimo nonlinear systems with time-varying delays, *International Journal of Robust and Nonlinear Control* **24**, 18, pp. 3474–3490.

Yu, Y., Lu, G., Sun, C. Y., and Liu, H. (2015). Robust backstepping decentralized tracking control for a 3-dof helicopter, *Nonlinear Dynamics* **82**, 1-2, pp. 947–960.

Chapter 5
Distributed Event-Based Consensus Control of Multi-agent System with Matching Nonlinear Uncertainties

In practical systems, autonomous agents usually are equipped with digital microprocessors so that digital platforms are needed to coordinate the data acquisition, transmission and controller execution. To address this concern, one approach is sampled control such as Xie et al. (2009), in which the controllers are updated periodically. But with this method the controller updates continue periodically with the same frequency even after the control goals have been achieved with sufficient enough accuracy. To overcome this shortcoming, it is preferable to introduce event-triggered schemes. The idea of event driven schemes is produced in computing, in which an event is an action or occurrence recognized by software that may be handled by the software. Computer events can be generated or triggered by the system, by the user or in other ways. Software that changes its behavior in response to events is said to be event-driven. Nowadays, there are many results based on event-triggered schemes for multi-agent systems such as Tabuada (2007), Grune and Muller (2009), Heemels et al. (2008). In these results, controller updates depend on the prescribed events, which can reduce the number of controller updates greatly, avoid unnecessary bandwidth and energy consumption. Meanwhile, event-triggered scheme has been addressed in multi-agent systems due to its advantages.

The event-triggered controllers can be developed with continuous or dis-continuous communication among agents. In early works, event-triggered controllers with continuous communication are assumed to achieve cooperative control. In Mazo and Tabuada (2011), Zhang et al. (2014), Fan et al. (2013) the controller updates depend on the certain measurement errors and the norm of the agents' states, in which continuous communication with neighboring agents is necessary. As we all know unnecessary communication can result in the waste of bandwidth and energy. Therefore, it is preferable to develop event-triggered schemes with dis-continuous communication, which can reduce the transmission frequency and save energy resources greatly. This problem is meaningful but challenging. Recently, many results have been obtained. For single-integrator multi-agent systems, in Dimarogonas et al. (2012), Liu et al. (2017) and Fan and Yang (2016) state-dependent triggering strate-

Q. Wang et al., *Distributed Cooperative Control and Optimization for Multi-Agent Systems*, https://doi.org/10.1007/978-981-96-0950-5_5

gies are proposed to avoid continuous measurements. In Garcia et al. (2016), time-dependent triggering functions are designed for second-order multi-agent systems. For general linear systems, time-dependent triggering functions are chosen to avoid continuous communication between neighboring agents in Yang et al. (2016) and Wu et al. (2017). In Zhu and Jiang (2015), triggering functions combined with states and time are designed to achieve leader-follower tracking for multi-agent system with general linear dynamics. In Meng and Chen (2013), with the quadratic Lyapunov function an event triggering scheme is proposed for multi-integrator multi-agent systems over fixed/switching, undirected but connected network topologies. In Hu and Liu (2017), the cooperative output regulation problem of heterogeneous MASs is solved and a novel event-triggered control mechanism is introduced to avoid continuous monitoring of measurement error. In Zhang and Han (2014), a novel output-based discrete event-triggered scheme is introduced to choose the necessary sampled-data packets transmitted through the communication network for control protocols design. In our paper, we also consider the event-triggered schemes for high-order multi-agent systems with dis-continuous communication.

For nonlinear multi-agent systems, the development of event-triggered schemes with dis-continuous communication remains meaningful and challenging. In practice, it is necessary to consider the nonlinear uncertainties due to uncertainties including external disturbances, varying parameters, noise, delay, missing measurements and so on. In the last several decades, recursive state estimation (Du et al. 2016) and optimal filter design (Hu et al. 2016) for networked stochastic systems subject to state delay and missing measurements were solved. In Mu et al. (2017a, b), data-driven adaptive tracking control approach is proposed for a class of continuous-time nonlinear systems. Though there are lots of results about nonlinear systems, to the authors' knowledge, there exist a few results for nonlinear multi-agent systems with dis-continuous event-triggered strategies. In Liu et al. (2017), Li et al. (2016), Liuzza et al. (2016), single-integrator multi-agent systems with Lipschitz nonlinear uncertainties are considered and dis-continuous event-triggered schemes are proposed to achieve consensus. In Li et al. (2016), the second-order dynamics with nonlinear uncertainties are considered and state-dependent event-triggered mechanism without continuous communication among agents are proposed. However, the mentioned works above are all about low-order nonlinear systems and few works involve event-triggered schemes with dis-continuous communication for high-order nonlinear multi-agent systems, which motivates our research. In this chapter, we will consider the event-triggered consensus problem for high-order multi-agent systems with norm-bounded uncertainties. Encouraged by the adaptive method in Li et al. (2014), we propose the event-based adaptive law to restrain the uncertainties and achieve consensus. With the proposed event-triggered adaptive controller, the main contributions are as the following:

(1) The multi-agent systems described by high-order models with norm-bounded nonlinear uncertainties are considered;
(2) With the proposed event-triggered schemes, continuous controller updates and communication among agents are avoided;
(3) There exists no Zeno behavior;
(4) The adaptive controller can guarantee the boundedness of consensus errors; besides, all the agents can be proved to converge to a small adjustable bounded set exponentially.

5.1 Mathematical Preliminaries

5.1.1 Notations

In the following sections, I_N denotes the N-dimensional identity matrix. $\mathbf{0}$ and $\mathbf{1}_N$ represent the column vectors with all elements being 0 and 1, respectively. $A \cdot B$ and $A \otimes B$ denote the product and Kronecker product of matrices A and B, respectively. For the real symmetric matrix A, $\lambda_{max}(A)$ and $\lambda_{min}(A)$ denote its maximum and minimum eigenvalues, respectively, and $A > 0(A < 0)$ if A is a positive (negative) definite matrix. For vectors, $\|\cdot\|$ denotes the Euclidean norm; for matrices, it denotes the induced 2-norm.

5.1.2 Graph Theory

For a graph $\mathcal{G}$ with n vertices, denoting the graph $\mathcal{G} = \{\mathcal{V}, \mathcal{E}\}$. $\mathcal{V} = \{v_1, v_2, \ldots, v_n\}$ represents the nodes, and $\mathcal{E} = \{(v_i, v_j)|v_i, v_j \in \mathcal{V}\}$ is the set of edges. $\mathcal{G}$ is called undirected if $(v_i, v_j) \in \mathcal{E} \Leftrightarrow (v_j, v_i) \in \mathcal{E}$. Meanwhile, if there is an edge between v_i and v_j, then it is said that node v_i and node v_j are adjacent. The adjacency matrix $\mathcal{A} = \left[a_{ij}\right] \in R^{n\times n}$ associated with $\mathcal{G}$ is defined as $a_{ij} = 1$ if $(v_j, v_i) \in \mathcal{E}$. For each agent v_i, N_i is the cardinality of its neighbor set and $N_{min} = \min\{N_i\}$. The Laplacian matrix $L = \left[L_{ij}\right] \in R^{n\times n}$ is defined as

$$\begin{cases} L_{ii} = \sum\limits_{j=1, j\neq i}^{n} a_{ij}, i = j, \\ L_{ij} = -a_{ij}, i \neq j. \end{cases}$$

Obviously, for an undirected graph, L is symmetric and positive semi-finite, i.e.$L^T = L \geq 0$. The row sum of L is zero; thus, the corresponding eigenvector of eigenvalue $\lambda_1(\mathcal{G}) = 0$ is $\mathbf{1}$,i.e.,$L\mathbf{1} = \mathbf{0}$. For a connected graph, there exists exactly one zero eigenvalue for L, and the eigenvalues are sorted in ascending order $0 = \lambda_1(\mathcal{G}) < \lambda_2(\mathcal{G}) \leq \cdots \leq \lambda_N(\mathcal{G})$.

5.2 Problem Formulation

In the paper, we consider a general class of N agents with heterogeneous uncertainties. The dynamics of each agent are described by

$$\dot{x}_i(t) = Ax_i(t) + B\left(u_i(t) + f_i\left(t, x_i(t)\right)\right), \tag{5.1}$$

where $i = 1, \ldots, N$, $x_i(t) \in R^n$ and $u_i(t) \in R^p$ respectively denote the states and control inputs of the i-th agent. $f_i(t, x_i(t))$ denote the heterogeneous uncertainties of i-th agent satisfying the following assumption. $A \in R^{n\times n}$ and $B \in R^{n\times p}$ are constant matrices.

Assumption 5.1 Graph $\mathcal{G}$ is connected and undirected.

Assumption 5.2 There are unknown positive constants ε_i satisfying that $\|f_i(x_i, t)\| \leq \varepsilon_i\|x_i\|, i = 1, \ldots, N$.

Remark 5.1 The uncertainties $\|f_i(x_i, t)\|$ associated with the i-th agent include external disturbances and parameter uncertainties, which satisfy the standard matching condition. The constants ε_i in Assumption 5.2 are not required to be given in the distributed consensus protocols to be developed, which is more general than the Lipschitz conditions.

Remark 5.2 The aim of the following section is to deal with the consensus problem for the multi-agent systems in (1), i.e., to propose distributed consensus protocols so that $\lim_{t\to\infty} \|x_i(t) - \sum_{j=1}^{N} \xi_j x_j(t)\| = 0$ $(\forall i, j = 1, \ldots, N)$, where $\xi = [\xi_1, \xi_2, \ldots, \xi_N]^T$ is the nonnegative left eigenvector of L associated with eigenvalue zero satisfying $\sum_{j=1}^{N} \xi_j = 1$ (Yu et al. 2011).

Lemma 5.1 (Ren and Beard (2005)) *For the graph $\mathcal{G}$ satisfying Assumption 1, 0 is an eigenvalue of L with the corresponding right eigenvector* $\mathbf{1}$ *and all other nonzero eigenvalues have positive real parts. Moreover, if and only if $\mathcal{G}$ has a directed spanning tree, 0 is a simple eigenvalue of L.*

In this chapter, we will research the event-triggered consensus control problem of system (5.1) and design a triggering rule to determine when i-th has to trigger and broadcast its new state to its neighbors so that all the agents can exponentially converge to a bounded set.

5.3 Consensus Analysis

The proposed distributed adaptive event-triggered protocol for each agent i is listed as follows:

$$
\begin{cases}
u_i(t) = cK \sum_{j=1}^{N} a_{ij} \left(x_i(t_k^i) - x_j(t_{k'}^j) \right) \\
\qquad + r_i \left(K \sum_{j=1}^{N} a_{ij} \left(x_i(t_k^i) - x_j(t_{k'}^j) \right) \right), \\
\dot{\bar{\varepsilon}}_i(t) = \epsilon_i \Big[- \psi_i \bar{\varepsilon}_i(t) + (1 + \beta_1) \Big\| K \sum_{j=1}^{N} a_{ij} \big(x_i(t_k^i) \\
\qquad - x_j(t_{k'}^j) \big) \Big\| \, \| x_i(t_k^i) \| \Big], i = 1, \ldots, N,
\end{cases}
\tag{5.2}
$$

where $t \in [t_k^i, t_{k+1}^i)$ and $\{t_k^i\}$, $k = 0, 1, \ldots$ denote the triggering time instants of i-th agent. $t_{k'}^j = \arg\min_{l \in \mathbb{N}: t \geq t_l^j} \{t - t_l^j\}$ is the last triggering time of agent j. Therefore, each agent updates its own control law at its own triggering times $t_0^i, t_1^i, \ldots$, as well as at the triggering times of its neighbors $t_0^j, t_1^j, \ldots$, and j is the neighbor of agent i. $\bar{\varepsilon}_i(t)$ is the adaptive gains corresponding to the i-th agent, a_{ij} is the (i, j)-th entry of the adjacency matrix $\mathcal{A}$ corresponding to $\mathcal{G}$, $K \in R^{p \times n}$ and $\Gamma \in R^{n \times n}$ are the feedback gain matrices, ϵ_i are positive scalars, ψ_i are small positive constants given by the designer, the nonlinear functions $r_i(\cdot)$ are designed as the following: for $w \in R^n$,

$$
r_i(w) = \begin{cases}
\dfrac{2w\bar{\varepsilon}_i \|x_i(t_k^i)\|}{(1 - \beta_1)\|w\|} & \text{if } \bar{\varepsilon}_i \|x_i(t_k^i)\| \|w\| > \kappa_i \\
\dfrac{2w\bar{\varepsilon}_i^2 \|x_i(t_k^i)\|^2}{(1 - \beta_1)\kappa_i} & \text{if } \bar{\varepsilon}_i \|x_i(t_k^i)\| \|w\| \leq \kappa_i
\end{cases}
\tag{5.3}
$$

where κ_i are small positive constants.

Choose the measurement errors $e_i(t) = x_i(t_k^i) - x_i(t)$, $t \in [t_k^i, t_{k+1}^i)$ for the ith agent. Considering that each agent can only gain the sampling states of its neighbors, the event is computed relying on its local information that is available to each agent. We design the triggering function:

$$
t_{k+1}^i = inf\{t : t > t_k^i, E_i(t) = 0\}
\tag{5.4}
$$

where $E_i(t) = \|e_i(t)\| - \beta_1/N_i \left\| \sum_{j \in N_i} \left(x_i(t_k^i) - x_j(t_{k'}^j) \right) \right\|$ and $\beta_1 > 0$.

Remark 5.3 In (5.4), $e_i(t)$ is the defined measurement error. When $e_i(t)$ becomes large so that $E_i(t)$ defined in (5.4) equals to zero, the actuator works and the controller $u_i(t)$ is updated. Otherwise, the controller $u_i(t)$ remains unchanged.

With the controllers above, each agent i can monitor its own state continuously. As the measurement error exceeds a certain threshold, implying $E_i(t) \geqslant 0$, an event is triggered for agent i. At the event-triggered instant, agent i will update its own controller with its own states and the neighbors' states stored. Meanwhile, agent i broadcasts its current state to its neighbors and measurement error $e_i(t)$ will be set to zero automatically.

When the multi-agent system achieves consensus, the trajectories of all the agents can converge to a time-varying weighed average $\bar{x}(t) = \sum_{j=1}^{N} \xi_j x_j(t)$, where $\xi = [\xi_1, \xi_2, \ldots, \xi_N]^T$ is the nonnegative left eigenvector of L corresponding to eigenvalue zero such that $\sum_{j=1}^{N} \xi_j = 1$. The following is to analyze whether the final consensus $\bar{x}(t)$ can still be achieved with the event-based scheme.

Define $\hat{x}_i(t) = x_i(t) - \bar{x}(t)$ and we have the fact that $\xi^T L = 0$, $\sum_{j=1}^{N} L_{ij} = 0$ and $\sum_{j=1}^{N} L_{ij}\bar{x}(t) = 0$. Then substituting (5.2) into (5.1), one can obtain the following error dynamics:

$$
\begin{aligned}
\dot{\hat{x}}_i(t) &= Ax_i(t) + cBK\sum_{j=1}^{N} a_{ij}\left(x_j(t^j_{k'}) - x_i(t^i_k)\right) \\
&\quad + Br_i\left(K\sum_{j=1}^{N} a_{ij}\left(x_j(t^j_{k'}) - x_i(t^i_k)\right)\right) \\
&\quad + Bf_i(x_i(t), t) - \sum_{j=1}^{N}\xi_j \dot{x}_j(t) \\
&= Ax_i(t) + Bf_i(x_i(t), t) \\
&\quad + cBK\sum_{j=1}^{N} a_{ij}\left(x_j(t^j_{k'}) - x_i(t^i_k)\right) \\
&\quad + Br_i\left(K\sum_{j=1}^{N} a_{ij}\left(x_j(t^j_{k'}) - x_i(t^i_k)\right)\right) - \\
&\quad \sum_{j=1}^{N}\xi_j\Bigg[Ax_j(t) + cBK\sum_{l=1}^{N} a_{jl}\left(x_l(t^l_{k''}) - x_j(t^j_{k'})\right) \\
&\quad + Br_j\left(K\sum_{l=1}^{N} a_{jl}\left(x_l(t^l_{k''}) - x_j(t^j_{k'})\right)\right) \\
&\quad + Bf_j(x_j(t), t)\Bigg]
\end{aligned}
$$

$$
\begin{aligned}
=&Ax_i(t) - cBK\sum_{j=1}^{N} L_{ij}\left(x_j(t)+e_j(t)\right)\\
&+ Br_i\left(-K\sum_{j=1}^{N} L_{ij}\left(x_j(t)+e_j(t)\right)\right) + Bf_i(x_i(t),t)\\
&-\sum_{j=1}^{N}\xi_j\Bigg[Ax_j(t) + Bf_j\left(x_j(t),t\right)\\
&- cBK\sum_{l=1}^{N} L_{jl}\left(x_l(t)+e_l(t)\right)\\
&+ Br_j\left(-K\sum_{l=1}^{N} L_{jl}\left(x_l(t)+e_l(t)\right)\right)\Bigg].
\end{aligned}
\tag{5.5}
$$

Define that $\hat{x}(t) = \left[\hat{x}_1^T(t), \hat{x}_2^T(t), \ldots, \hat{x}_N^T(t)\right]^T$, and it is easy to derive that $\hat{x}(t) = ((\mathbf{I}_N - \mathbf{1}_N\xi^T)\otimes I_n)x(t)$. From (5.5), the system can be further written as follows:

$$
\begin{aligned}
\dot{\hat{x}}(t) =&(\mathbf{I}_N\otimes A)x(t) - (cL\otimes BK)\left(x(t)+e(t)\right)\\
&+ (\mathbf{I}_N\otimes B)F(x(t),t) + (\mathbf{I}_N\otimes B)R(\hat{x}(t),t) -\\
&(\mathbf{1}_N\xi^T\otimes A)x(t) + \left(c\mathbf{1}_N\xi^T L\otimes BK\right)\left(x(t)+e(t)\right)\\
&- (\mathbf{1}_N\xi^T\otimes B)F(x(t),t) - (\mathbf{1}_N\xi^T\otimes B)R(\hat{x}(t),t)\\
=&\left((\mathbf{I}_N - \mathbf{1}_N\xi^T)\otimes A\right)x(t)\\
&- \left(c(\mathbf{I}_N - \mathbf{1}_N\xi^T)L\otimes BK\right)\left(x(t)+e(t)\right)\\
&+ ((\mathbf{I}_N - \mathbf{1}_N\xi^T)\otimes B)F(x(t),t)\\
&+ ((\mathbf{I}_N - \mathbf{1}_N\xi^T)\otimes B)R(\hat{x}(t),t)\\
=&\left(\mathbf{I}_N\otimes A - c(\mathbf{I}_N - \mathbf{1}_N\xi^T)L\otimes BK\right)\hat{x}(t)\\
&- \left(c(\mathbf{I}_N - \mathbf{1}_N\xi^T)L\otimes BK\right)e(t)\\
&+ \left((\mathbf{I}_N - \mathbf{1}_N\xi^T)\otimes B\right)F(x(t),t)\\
&+ \left((\mathbf{I}_N - \mathbf{1}_N\xi^T)\otimes B\right)R(\hat{x}(t),t)
\end{aligned}
\tag{5.6}
$$

where $F(x(t),t) \triangleq \begin{bmatrix} f_1(x_1(t),t)\\ \vdots \\ f_N(x_N(t),t)\end{bmatrix}$ and $R(\hat{x}(t),t) \triangleq \begin{bmatrix} r_1\left(-K\sum_{j=1}^{N} L_{1j}\left(x_j(t)+e_j(t)\right)\right)\\ \vdots \\ r_N\left(-K\sum_{j=1}^{N} L_{Nj}\left(x_j(t)+e_j(t)\right)\right)\end{bmatrix}$.

Theorem 5.1 *Assume that (A, B) in the systems (5.1) is stabilizable and the Assumption 1 and 2 holds. The multi-agent systems (5.1) can achieve consensus with the control strategies (5.2) and the triggering function defined by (5.4). The feedback gain matrices of the control protocols are given by $K = -B^T P$ and $P > 0$ is the unique solution to the algebraic Riccati equation (ARE):*

$$A^T P + PA - PBB^T P + I = 0. \tag{5.7}$$

Under the distributed event-triggered sampling function, the global consensus error $\hat{x}$ exponentially converges to the set

$$\mathcal{D} = \left\{ \hat{x} : \|\hat{x}\|^2 \leq \frac{1}{\delta \lambda_2 \lambda_{min}(P)} \sum_{i=1}^{N} \left(\psi_i \varepsilon_i^2 + \frac{1+\beta_1}{2} \kappa_i \right) \right\} \tag{5.8}$$

where $0 < \delta \leq \min\{\alpha, \varphi_i \tau_i, \psi_i \epsilon_i\}, \alpha = \dfrac{1}{\lambda_{max}(P)}$, $c > 1/\lambda_2(1-\rho^2)$, $0 < \beta_1 < N_{min}/(N + N_{min})$ *and* $\rho = \beta_1(N + N_{min})/[N_{min} - \beta_1(N + N_{min})] > 0$.

Furthermore, the multi-agent systems (5.1) *with the control strategies* (5.2) *exhibits no Zeno behavior.*

Proof Choosing the following Lyapunov function:

$$V = 1/2\hat{x}^T (L \otimes P)\hat{x} + \sum_{i=1}^{N} \frac{\Delta \varepsilon_i^2}{2\epsilon_i} \tag{5.9}$$

where $\Delta \varepsilon_i = \bar{\varepsilon}_i - \varepsilon_i$.

Substituting (5.2) and (5.6) into (5.9), the time derivative of V can be obtained as

$$\begin{aligned}
\dot{V} =& \hat{x}^T(t)(L \otimes P)\Big[(\mathbf{I}_N \otimes A + c(\mathbf{I}_N - \mathbf{1}_N \xi^T)L \otimes BK)\hat{x}(t) \\
&+ (c(\mathbf{I}_N - \mathbf{1}_N \xi^T)L \otimes BK)e(t)\Big] \\
&+ \hat{x}^T(t)(L \otimes P)\Big[((\mathbf{I}_N - \mathbf{1}_N \xi^T) \otimes B)F(x(t), t) \\
&+ ((\mathbf{I}_N - \mathbf{1}_N \xi^T) \otimes B)R(\hat{x}(t), t)\Big] + \sum_{i=1}^{N} \Delta \varepsilon_i \Big[- \psi_i \bar{\varepsilon}_i(t) \\
&+ (1+\beta_1) \left\| K \sum_{j=1}^{N} a_{ij} \left(x_i(t_i^k) - x_j(t_j^{k'}) \right) \right\| \|x_i(t_i^k)\| \Big] \\
=& \hat{x}^T(t)(L \otimes PA + cLL \otimes PBK)\hat{x}(t) \\
&+ \hat{x}^T(t)(L \otimes PB)(F(x(t), t) + R(\hat{x}(t), t))
\end{aligned}$$

$$
\begin{aligned}
&+ \hat{x}^T(t)(cLL \otimes PBK)e(t) \\
&+ \sum_{i=1}^{N} \Delta\varepsilon_i \Bigg[- \psi_i \bar{\varepsilon}_i(t) + (1+\beta_1) \\
&\cdot \left\| K \sum_{j=1}^{N} a_{ij} \left(x_i(t_i^k) - x_j(t_j^{k'}) \right) \right\| \|x_i(t_i^k)\| \Bigg].
\end{aligned} \tag{5.10}
$$

From the Assumption 5.2, it is easy to obtain that

$$
\begin{aligned}
&\hat{x}^T(t)(L \otimes PB)(F(x(t),t)) \\
&\leq \sum_{i=1}^{N} \varepsilon_i \|x_i\| \left\| B^T P \sum_{j=1}^{N} L_{ij} \hat{x}_j \right\| \\
&= \sum_{i=1}^{N} \varepsilon_i \|x_i\| \left\| B^T P \sum_{j=1}^{N} L_{ij} \left(x_j(t) + \sum_{l=1}^{N} \xi_l x_l(t) \right) \right\| \\
&= \sum_{i=1}^{N} \varepsilon_i \|x_i\| \left\| B^T P \sum_{j=1}^{N} L_{ij} x_j(t) \right\|.
\end{aligned} \tag{5.11}
$$

Due to the defined nonlinear functions $r_i(\cdot)$ in (5.3), our analysis can be divided into two cases as follows:

Case (I): $\bar{\varepsilon}_i \|x_i\| \left\| K \sum_{j=1}^{N} L_{ij} \left(x_j(t) + e_j(t) \right) \right\| > \kappa_i$

$$
\begin{aligned}
&\hat{x}^T(t)(L \otimes PB)(R(x(t),t)) \\
&\leq \sum_{i=1}^{N} \frac{2(\bar{\varepsilon}_i \|x_i\|) K \sum_{j=1}^{N} L_{ij} \left(x_j(t) + e_j(t) \right)}{(1-\beta_1) \left\| K \sum_{j=1}^{N} L_{ij} \left(x_j(t) + e_j(t) \right) \right\|} \cdot \left(B^T P \sum_{j=1}^{N} L_{ij} \hat{x}_j(t) \right) \\
&= \sum_{i=1}^{N} \frac{-2(\bar{\varepsilon}_i \|x_i\|) B^T P \sum_{j=1}^{N} L_{ij} \left(x_j(t) + e_j(t) \right)}{(1-\beta_1) \left\| B^T P \sum_{j=1}^{N} L_{ij} \left(x_j(t) + e_j(t) \right) \right\|} \\
&\cdot \left(B^T P \sum_{j=1}^{N} L_{ij} \left(x_j(t) + \sum_{l=1}^{N} \xi_l x_l(t) \right) \right) \\
&= - \sum_{i=1}^{N} \frac{2(\bar{\varepsilon}_i \|x_i\|) B^T P \sum_{j=1}^{N} L_{ij} \left(x_j(t) + e_j(t) \right)}{(1-\beta_1) \left\| B^T P \sum_{j=1}^{N} L_{ij} \left(x_j(t) + e_j(t) \right) \right\|} \cdot \left(B^T P \sum_{j=1}^{N} L_{ij} x_j(t) \right)
\end{aligned}
$$

$$
\begin{aligned}
=&\sum_{i=1}^{N}\frac{-2}{1-\beta_1}(\bar{\varepsilon}_i\|x_i\|)\left\|B^TP\sum_{j=1}^{N}L_{ij}\left(x_j(t)+e_j(t)\right)\right\| \\
&+\sum_{i=1}^{N}\frac{2(\bar{\varepsilon}_i\|x_i\|)B^TP\sum_{j=1}^{N}L_{ij}\left(x_j(t)+e_j(t)\right)}{(1-\beta_1)\left\|B^TP\sum_{j=1}^{N}L_{ij}\left(x_j(t)+e_j(t)\right)\right\|}\cdot\left(B^TP\sum_{j=1}^{N}L_{ij}e_j(t)\right)
\end{aligned}
\tag{5.12}
$$

From the fact that $2ab \le a^2+b^2$ and the defined triggering function (5.4), one can further derive that

$$
\begin{aligned}
&\sum_{i=1}^{N}\frac{2(\bar{\varepsilon}_i\|x_i\|)B^TP\sum_{j=1}^{N}L_{ij}\left(x_j(t)+e_j(t)\right)}{(1-\beta_1)\left\|B^TP\sum_{j=1}^{N}L_{ij}\left(x_j(t)+e_j(t)\right)\right\|}\cdot\left(B^TP\sum_{j=1}^{N}L_{ij}e_j(t)\right) \\
\le&\frac{1}{1-\beta_1}\sum_{i=1}^{N}(\bar{\varepsilon}_i\|x_i\|)\left\|B^TP\sum_{j=1}^{N}L_{ij}\left(x_j(t)+e_j(t)\right)\right\| \\
&+\frac{1}{1-\beta_1}\sum_{i=1}^{N}\frac{(\bar{\varepsilon}_i\|x_i\|)\left\|B^TP\sum_{j=1}^{N}L_{ij}e_j(t)\right\|^2}{\left\|B^TP\sum_{j=1}^{N}L_{ij}\left(x_j(t)+e_j(t)\right)\right\|} \\
\le&\frac{1}{1-\beta_1}\sum_{i=1}^{N}(\bar{\varepsilon}_i\|x_i\|)\left\|B^TP\sum_{j=1}^{N}L_{ij}\left(x_j(t)+e_j(t)\right)\right\| \\
&+\frac{\beta_1^2}{1-\beta_1}\sum_{i=1}^{N}\frac{(\bar{\varepsilon}_i\|x_i\|)\left\|B^TP\sum_{j=1}^{N}L_{ij}\left(x_j(t)+e_j(t)\right)\right\|^2}{\left\|B^TP\sum_{j=1}^{N}L_{ij}\left(x_j(t)+e_j(t)\right)\right\|}
\end{aligned}
\tag{5.13}
$$

Substituting (5.13) into (5.12), one can have

$$
\begin{aligned}
&\hat{x}^T(t)(L\otimes PB)(R(x(t),t)) \\
\le&-\frac{1}{1-\beta_1}\sum_{i=1}^{N}(\bar{\varepsilon}_i\|x_i\|)\left\|B^TP\sum_{j=1}^{N}L_{ij}\left(x_j(t)+e_j(t)\right)\right\| \\
&+\frac{\beta_1^2}{1-\beta_1}\sum_{i=1}^{N}\frac{(\bar{\varepsilon}_i\|x_i\|)\left\|B^TP\sum_{j=1}^{N}L_{ij}\left(x_j(t)+e_j(t)\right)\right\|^2}{\left\|B^TP\sum_{j=1}^{N}L_{ij}\left(x_j(t)+e_j(t)\right)\right\|} \\
\le&-(1+\beta_1)\sum_{i=1}^{N}\bar{\varepsilon}_i\|x_i\|\left\|B^TP\sum_{j=1}^{N}L_{ij}\left(x_j(t)+e_j(t)\right)\right\|
\end{aligned}
\tag{5.14}
$$

Substituting (5.11) and (5.14) into (5.10), we can derive the following

$$
\begin{aligned}
\dot{V} \leq & \hat{x}^T(t)(L \otimes PA - cLL \otimes PBB^T P)\hat{x}(t) + \hat{x}^T(t) \\
& (cLL \otimes PBK)e(t) + \sum_{i=1}^{N} \varepsilon_i \|x_i\| \left\| B^T P \sum_{j=1}^{N} L_{ij} x_j \right\| \\
& - (1+\beta_1) \sum_{i=1}^{N} \bar{\varepsilon}_i \|x_i\| \left\| B^T P \sum_{j=1}^{N} L_{ij} \left(x_j(t) + e_j(t) \right) \right\| \\
& - \sum_{i=1}^{N} \psi_i \Delta\varepsilon_i (\Delta\varepsilon_i + \varepsilon_i) \\
& + (1+\beta_1) \sum_{i=1}^{N} \Delta\varepsilon_i \|x_i\| \left\| B^T P \sum_{j=1}^{N} L_{ij} \left(x_j + e_j \right) \right\| . \qquad (5.15)
\end{aligned}
$$

From the defined triggering function (5.4), one can further derive that

$$
\begin{aligned}
& \sum_{i=1}^{N} \varepsilon_i \|x_i\| \left\| B^T P \sum_{j=1}^{N} L_{ij} x_j \right\| \\
\leq & \sum_{i=1}^{N} \varepsilon_i \|x_i\| \left[\left\| B^T P \sum_{j=1}^{N} L_{ij}(x_j + e_j) \right\| + \left\| B^T P \sum_{j=1}^{N} L_{ij} e_j \right\| \right] \\
= & \sum_{i=1}^{N} (\varepsilon_i \|x_i\|) \left\| B^T P \sum_{j=1}^{N} L_{ij}(x_j + e_j) \right\| \\
& + \sum_{i=1}^{N} (\varepsilon_i \|x_i\|) \left\| B^T P \sum_{j=1}^{N} a_{ij} \left(e_i - e_j \right) \right\| \\
\leq & (1+\beta_1) \sum_{i=1}^{N} (\varepsilon_i \|x_i\|) \left\| B^T P \sum_{j=1}^{N} L_{ij}(x_j + e_j) \right\| . \qquad (5.16)
\end{aligned}
$$

From (5.15) and (5.16), one can derive

$$
\begin{aligned}
\dot{V} \leq & \hat{x}^T(t)(L \otimes PA - cLL \otimes PBB^T P)\hat{x}(t) \\
& + \hat{x}^T(t)(cLL \otimes PBK)e(t) + (1+\beta_1)
\end{aligned}
$$

$$
\begin{aligned}
&\cdot \sum_{i=1}^{N}(\varepsilon_i \|x_i\|) \left\| B^T P \sum_{j=1}^{N} L_{ij} \left(x_j + e_j\right) \right\| - (1 \\
&+ \beta_1) \sum_{i=1}^{N}(\bar{\varepsilon}_i \|x_i\|) \left\| B^T P \sum_{j=1}^{N} L_{ij} \left(x_j(t) + e_j(t)\right) \right\| \\
&- \sum_{i=1}^{N} \psi_i \Delta\varepsilon_i \left(\Delta\varepsilon_i + \varepsilon_i\right) \\
&+ (1+\beta_1) \sum_{i=1}^{N} \Delta\varepsilon_i \|x_i\| \left\| B^T P \sum_{j=1}^{N} L_{ij} \left(x_j + e_j\right) \right\| \\
\leq\ & \hat{x}^T(t)(L \otimes PA - cLL \otimes PBB^T P)\hat{x}(t) \\
&+ \hat{x}^T(t)(cLL \otimes PBK)e(t) - \sum_{i=1}^{N} \psi_i \Delta\varepsilon_i \left(\Delta\varepsilon_i + \varepsilon_i\right) \\
\leq\ & -\delta V + \delta V - \sum_{i=1}^{N} \psi_i \Delta\varepsilon_i \left(\Delta\varepsilon_i + \varepsilon_i\right) \\
&+ \frac{1}{2}\hat{x}^T \left(L \otimes (PA + A^T P) - cLL \otimes PBB^T P\right) \hat{x} \\
&+ \frac{1}{2} e^T(t)(cLL \otimes PBB^T P)e(t) \\
\leq & -\delta V + \frac{1}{2}\hat{x}^T(t)\Big(L \otimes (PA + A^T P) \\
&- cLL \otimes PBB^T P + \delta L \otimes P\Big)\hat{x}(t) \\
&+ \frac{1}{2} e^T(t)(cLL \otimes PBB^T P)e(t) \\
&- \frac{1}{2}\sum_{i=1}^{N}\left[\left(\psi_i - \frac{\delta}{\epsilon_i}\right)\Delta\varepsilon_i^2\right] + \frac{1}{2}\sum_{i=1}^{N}\psi_i\varepsilon_i^2.
\end{aligned}
\tag{5.17}
$$

Case (II): $\bar{\varepsilon}_i \|x_i\| \left\| K \sum_{j=1}^{N} L_{ij} \left(x_j(t) + e_j(t)\right) \right\| \leq \kappa_i$

$$
\begin{aligned}
&\hat{x}^T(t)(L \otimes PB)\left(R(x(t), t)\right) \\
&\leq \sum_{i=1}^{N} \frac{2(\bar{\varepsilon}_i \|x_i\|)^2 K \sum_{j=1}^{N} L_{ij}\left(x_j(t) + e_j(t)\right)}{(1-\beta_1)\kappa_i} \cdot \left(B^T P \sum_{j=1}^{N} L_{ij}\hat{x}_j(t) \right) \\
&= -\sum_{i=1}^{N} \frac{2(\bar{\varepsilon}_i \|x_i\|)^2 B^T P \sum_{j=1}^{N} L_{ij}\left(x_j(t) + e_j(t)\right)}{(1-\beta_1)\kappa_i} \cdot \left(B^T P \sum_{j=1}^{N} L_{ij} x_j(t) \right)
\end{aligned}
$$

$$
\begin{aligned}
&= - \sum_{i=1}^{N} \frac{2(\bar{\varepsilon}_i \|x_i\|)^2 B^T P \sum_{j=1}^{N} L_{ij} \left(x_j(t) + e_j(t)\right)}{(1-\beta_1)\kappa_i} \\
&\quad \cdot \left(B^T P \sum_{j=1}^{N} L_{ij} \left(x_j(t) + e_j(t) - e_j(t)\right) \right) \\
&= - \sum_{i=1}^{N} \frac{2(\bar{\varepsilon}_i \|x_i\|)^2 \left\| B^T P \sum_{j=1}^{N} L_{ij} \left(x_j(t) + e_j(t)\right) \right\|^2}{(1-\beta_1)\kappa_i} \\
&\quad + \sum_{i=1}^{N} \frac{2(\bar{\varepsilon}_i \|x_i\|)^2 B^T P \sum_{j=1}^{N} L_{ij} \left(x_j(t) + e_j(t)\right)}{(1-\beta_1)\kappa_i} \cdot \left(B^T P \sum_{j=1}^{N} L_{ij} e_j(t) \right).
\end{aligned} \tag{5.18}
$$

From the fact that $2ab \le a^2 + b^2$ and the defined triggering function (5.4), inequality (5.18) can be further expressed as

$$
\begin{aligned}
&\hat{x}^T(t)(L \otimes PB)\left(R(x(t), t)\right) \\
&\le \frac{-1}{1-\beta_1} \sum_{i=1}^{N} \frac{(\bar{\varepsilon}_i \|x_i\|)^2 \left\| B^T P \sum_{j=1}^{N} L_{ij} \left(x_j(t) + e_j(t)\right) \right\|^2}{\kappa_i} \\
&\quad + \frac{1}{1-\beta_1} \sum_{i=1}^{N} \frac{(\bar{\varepsilon}_i \|x_i\|)^2 \left\| B^T P \sum_{j=1}^{N} L_{ij} e_j(t) \right\|^2}{\kappa_i} \\
&\le \frac{\beta_1^2 - 1}{1-\beta_1} \sum_{i=1}^{N} \frac{(\bar{\varepsilon}_i \|x_i\|)^2 \left\| B^T P \sum_{j=1}^{N} L_{ij} \left(x_j(t) + e_j(t)\right) \right\|^2}{\kappa_i} \\
&= (1+\beta_1) \sum_{i=1}^{N} \frac{(\bar{\varepsilon}_i \|x_i\|)^2 \left\| B^T P \sum_{j=1}^{N} L_{ij} \left(x_j + e_j\right) \right\|^2}{-\kappa_i}.
\end{aligned} \tag{5.19}
$$

We have the fact that
$-(1+\beta_1) \sum_{i=1}^{N} \frac{(\bar{\varepsilon}_i \|x_i\|)^2 \left\| B^T P \sum_{j=1}^{N} L_{ij} \left(x_j(t) + e_j(t)\right) \right\|^2}{\kappa_i} + (1+\beta_1)$
$\sum_{i=1}^{N} (\bar{\varepsilon}_i \|x_i\|) \left\| B^T P \sum_{j=1}^{N} L_{ij} \left(x_j(t) + e_j(t)\right) \right\| \le \frac{(1+\beta_1)\kappa_i}{4}$, then (5.18) can be further expressed as

$$
\begin{aligned}
&\hat{x}^T(t)(L \otimes PB)\,(R(x(t),t)) \\
\leq & -(1+\beta_1)\sum_{i=1}^{N}(\bar{\varepsilon}_i\|x_i\|)\left\|B^T P\sum_{j=1}^{N}L_{ij}\left(x_j(t)+e_j(t)\right)\right\| \\
&+\frac{(1+\beta_1)\kappa_i}{4}.
\end{aligned}
\tag{5.20}
$$

Substituting (5.11) and (5.20) into (5.10), one can derive that

$$
\begin{aligned}
\dot{V} \leq & -\delta V+\frac{1}{2}\hat{x}^T(t)(L\otimes(PA+A^TP)-cLL\otimes PBB^TP \\
&+\delta L\otimes P)\hat{x}(t)+\frac{1}{2}e^T(t)(cLL\otimes PBB^TP)e(t)-\frac{1}{2} \\
&\cdot\sum_{i=1}^{N}\left[\left(\psi_i-\frac{\delta}{\epsilon_i}\right)\Delta\varepsilon_i^2\right]+\frac{1}{2}\sum_{i=1}^{N}\left(\psi_i\varepsilon_i^2+\frac{1+\beta_i}{2}\kappa_i\right)
\end{aligned}
\tag{5.21}
$$

Furthermore, combining the fact $\frac{1}{2}\sum_{i=1}^{N}\frac{1+\beta_i}{2}\kappa_i > 0$ and (5.17), we can obtain the same inequality in case (I) as (5.21) in case (II). Therefore, the inequality (5.21) is what we can obtain from the both two cases.

For any $t \in [t_k^i, t_{k+1}^i)$, from the definition of the measurement error $e_i(t)$, it is easy to see $e_i(t)$ will be reset to zero at each sampling instant t_k^i. That implies the next sampling time instant t_{k+1}^i is just the moment when the error $e_i(t)$ touches the individual threshold $\beta_1\sum_{j\in N_i}\|x_i(t_k^i)-x_j(t_{k'}^j)\|$ for each agent i. Therefore, for all t, we have

$$
\|e_i(t)\| \leq (\beta_1/N_i)\sum_{j\in N_i}\left\|x_i(t_k^i)-x_j(t_{k'}^j)\right\|. \tag{5.22}
$$

According to the definitions $e_i(t)=x_i(t_k^i)-x_i(t)$ and $\hat{x}_i(t)=x_i(t)-\bar{x}(t)$, it follows that

$$
\begin{aligned}
\|e_i(t)\| \leq & \beta_1\|\hat{x}_i(t)\|+\beta_1/N_i\|\hat{x}(t)\| \\
&+N_i\beta_1/N_i\|e_i(t)\|+\beta_1\|e(t)\|.
\end{aligned}
\tag{5.23}
$$

Thus, we have

$$
\begin{aligned}
\|e(t)\| \leq & \beta_1\|\hat{x}(t)\|+\frac{\beta_1 N}{N_{min}}\|\hat{x}(t)\| \\
&+\beta_1\|e(t)\|+\frac{\beta_1 N}{N_{min}}\|e(t)\|.
\end{aligned}
\tag{5.24}
$$

With the triggering function (5.4), the inequality (5.24) can be further written as

$$\begin{aligned}\|e(t)\| &\le \frac{\beta_1(N+N_{min})}{N_{min}-\beta_1(N+N_{min})}\|\hat{x}(t)\| \\ &= \rho\|\hat{x}(t)\|\end{aligned} \tag{5.25}$$

where $\rho = \dfrac{\beta_1(N+N_{min})}{N_{min}-\beta_1(N+N_{min})} > 0$, $\beta_1 > 0$ and $\beta_1 < N_{min}/(N+N_{min})$.

If we choose $\delta < \min\left\{\dfrac{1}{\lambda_{max}(P)}, \varphi_i\tau_i, \psi_i\epsilon_i\right\}$ and $c > 1/\lambda_2(1-\rho^2)$ as given in the Theorem, from (5.25), one can derive

$$\begin{aligned}&\frac{1}{2}\hat{x}^T(L\otimes(PA+A^TP)-cLL\otimes PBB^TP+\delta L\otimes P)\hat{x} \\ &\quad+\frac{1}{2}e^T(t)(cLL\otimes PBB^TP)e(t) \\ &\le \frac{1}{2}\sum_{i=1}^{N}\hat{x}_i^T\left(PA+A^TP+\delta P-(1-\rho^2)c\lambda_i PBB^TP\right)\hat{x}_i \\ &\le \sum_{i=1}^{N}\hat{x}_i^T\left(PA+A^TP+I-(1-\rho^2)c\lambda_i PBB^TP\right)\hat{x}_i \\ &\le 0.\end{aligned} \tag{5.26}$$

Combining (5.21) and (5.26), we have

$$\dot{V} \le -\delta V + \frac{1}{2}\sum_{i=1}^{N}\left(\psi_i\varepsilon_i^2 + \frac{1+\beta_i}{2}\kappa_i\right). \tag{5.27}$$

Further, one can obtain from (5.27) that

$$\begin{aligned}V \le &\left[V(0) - \frac{1}{2\delta}\sum_{i=1}^{N}\left(\psi_i\varepsilon_i^2 + \frac{1+\beta_i}{2}\kappa_i\right)\right]\exp(-\delta t) \\ &+\frac{1}{2\delta}\sum_{i=1}^{N}\left(\psi_i\varepsilon_i^2 + \frac{1+\beta_i}{2}\kappa_i\right).\end{aligned} \tag{5.28}$$

Obviously, we can see that $\hat{x}$ can exponentially converge to the bounded set

$$\|\hat{x}\|^2 < \frac{1}{\delta\lambda_2\lambda_{min}(P)}\sum_{i=1}^{N}\left(\psi_i\varepsilon_i^2 + \frac{1+\beta_1}{2}\kappa_i\right)$$

with a convergence rate larger than $\exp(-\delta t)$.

Remark 5.4 Compared with the existing work, the triggering functions (5.4) proposed in this paper does not require continuous communication, which is meaningful in practical systems. Besides, the proposed adaptive controllers are effective in restraining the nonlinear uncertainties, which offers a new way to deal with nonlinear systems with event-triggered scheme.

Now we are in the position to state that the Zeno behavior can be excluded with the triggering function (5.4) for the multi-agent systems (5.1).

Similar to Hu et al. (2017), we choose a more sufficient condition to take place of the triggering function (5.4) as the following:

$$e_i(t) \le \frac{\beta_1}{N_i\sqrt{1+\beta_1^2}} \left\| \sum_{j=1}^{N} a_{ij} \left(x_i(t_{k+1}^i) - x_j(t_{k'+1}^j) \right) \right\|. \tag{5.29}$$

Define $\varpi_i = \dfrac{\beta_1}{N_i\sqrt{1+\beta_1^2}} \left\| \sum_{j=1}^{N} a_{ij} \left(x_i(t_{k+1}^i) - x_j(t_{k'+1}^j) \right) \right\|$ and the triggering occurs when (5.29) becomes an equality, which can guarantee $E_i(t) \le 0$ in (5.4).

Define $u_i(t_k^i) = cK \sum_{j \in N_i} a_{ij} \left(x_i(t_k^i) - x_j(t_{k'}^j) \right)$, when $t \in [t_k^i, t_{k+1}^i)$. Combining the fact that $\|x_i(t)\| \le \|x_i(t_k^i)\| + \|e_i(t)\|$, we have the following inequality:

$$\begin{aligned}
\|\dot{e}_i(t)\| \le& \|A\|\|x_i(t)\| + \|B\|\|u_i(t_k^i)\| + \|B\|\|f_i(t,x_i)\| \\
\le& \|A\|\|x_i(t_k^i)\| + \|A\|\|e_i(t)\| \\
&+ \varepsilon_i\|B\|\|x_i(t)\| + \|B\|\|u_i(t_k^i)\| \\
\le& (\|A\|+\gamma)\|x_i(t_k^i)\| + (\|A\|+\gamma)\|e_i(t)\| \\
&+ \|B\|\|u_i(t_k^i)\|.
\end{aligned} \tag{5.30}$$

where γ is a constant satisfying $\gamma = \max\limits_{i=1,\dots,N} \varepsilon_i\|B\|$. From (5.30), it is easy to obtain

$$\|e_i(t)\| \le \alpha_k^i \left(e^{(\|A\|+\gamma)(t-t_k^i)} - 1 \right) \tag{5.31}$$

where $\alpha_k^i = \dfrac{(\|A\|+\gamma)\|x_i(t_k^i)\| + \|B\|\|u_i(t_k^i)\|}{\|A\|+\gamma}$.

Using (5.29) and (5.31) gives that,

$$e_i(t_{k+1}^i) = \varpi_i < \alpha_k^i \left(e^{(\|A\|+\gamma)(t_{k+1}^i - t_k^i)} - 1 \right), \tag{5.32}$$

which yields

$$t_{k+1}^i - t_k^i > \frac{1}{\|A\|+\gamma} \ln \left(1 + \frac{\varpi_i}{\alpha_k^i} \right). \tag{5.33}$$

Zeno behavior can be described as the system making an infinite number of jumps in a finite amount of time. From (5.33), it is easy to see that the inter-event interval is strictly positive for agent i, which implies that agent i exhibits no Zeno behavior. The proof of the Theorem is thus completed.

5.4 Numerical Simulation

A numerical simulation is given to verify the effectiveness of the theoretical result in this section. There are five agents, which are denoted by $i = 1, 2, \ldots, 5$, respectively. The communication topology of the agents are given as Fig. 5.1.

In this example, we suppose there exist five agents for the multi-agent system. The dynamics of each agent is described by:

$$A = \begin{bmatrix} 0.2 & 1 & 0 \\ -0.1 & -0.5 & 0.1 \\ 0.1 & -0.1 & -0.2 \end{bmatrix}; \quad B = \begin{bmatrix} 1 \\ 0 \\ 1 \end{bmatrix};$$

and the nonlinear uncertainties $f_i(t)$ are described by $f_i(t) = 0.5\|x_i(t)\|$ respectively. According to the Theorem, we can obtain the control gains as the following:

$$K = \begin{bmatrix} -2.7712 & -3.1470 & -0.8334 \end{bmatrix}$$

The initial state of x_i is chosen as $x_1(0) = [0.1 \;\; 0.2 \;\; 0.3]^T$, $x_2(0) = [0.2 \;\; 0.4 \;\; 0.6]^T$, $x_3(0) = [0.3 \;\; 0.6 \;\; 0.9]^T$,and $x_4(0) = [0.4 \;\; 0.8 \;\; 1.2]^T$, $x_5(0) = [0.5 \;\; 1 \;\; 1.5]^T$, respectively. The parameters $\kappa = 0.3$ and $\beta = 0.02$.

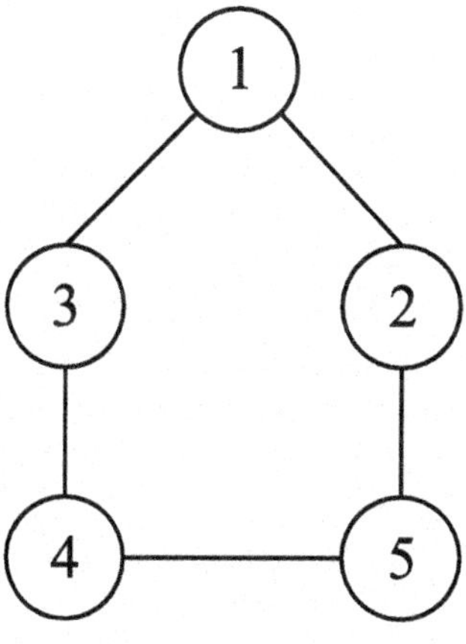

Fig. 5.1 The communication topology

Fig. 5.2 States x_{i1} of all agents($i = 1, \ldots, 5$)

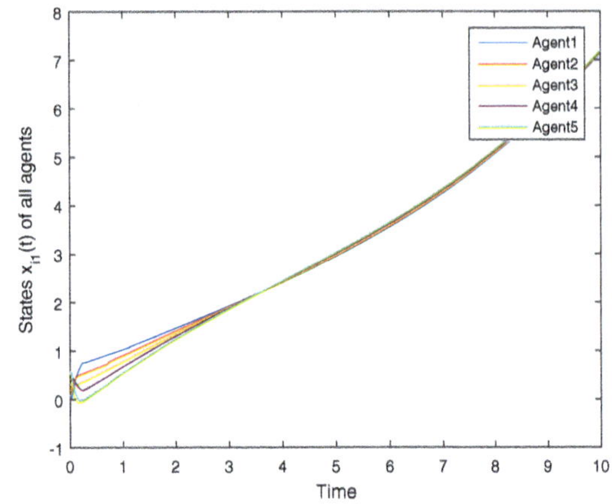

Fig. 5.3 States x_{i2} of all agents($i = 1, \ldots, 5$)

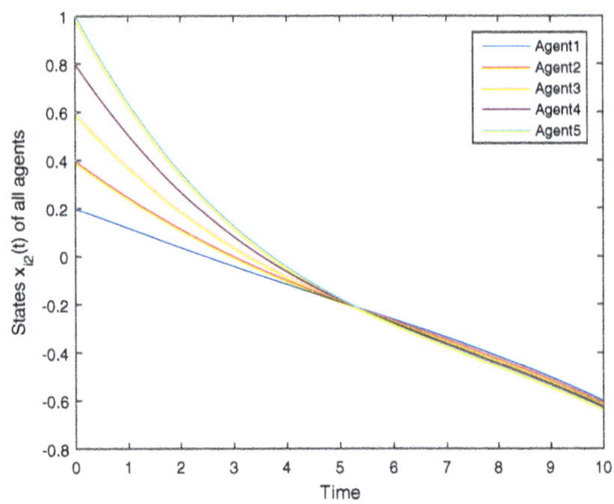

Figures 5.2, 5.3 and 5.4 are the states x_i of agent i and it is not difficult to find that the consensus can be reached. Figure 5.5 denotes the errors of the adaptive parameters $\bar{\varepsilon}_i - \varepsilon_i$ of each agent i and Fig. 5.6 denotes the input u_i, respectively.

In our simulations, the sampling times are chosen as 650, and the controller updates are listed as Table 5.1. It is easy to find that the times of controller updates are all less than sampling times, implying that the event-triggered scheme can reduce controller updates.

From the numeral simulation results, it is obvious that the consensus can be achieved and the controller updates are reduced in this example of the multi-agent system with the above nonlinear functions and parameters, which demonstrates the effectiveness of our methods.

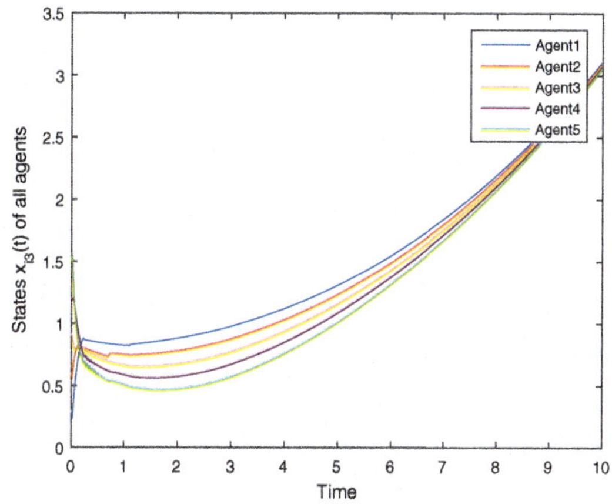

Fig. 5.4 States x_{i3} of all agents($i = 1, \ldots, 5$)

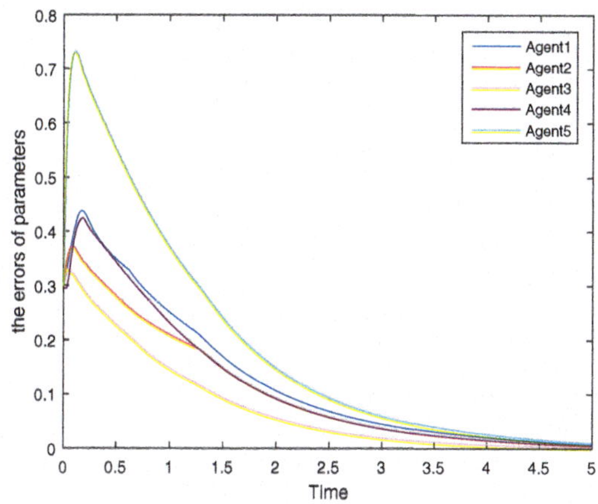

Fig. 5.5 The values of $\bar{\varepsilon}_i - \varepsilon_i (i = 1, \ldots, 5)$

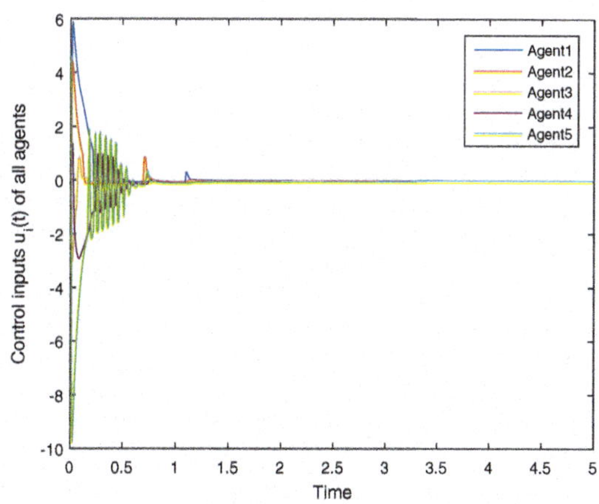

Fig. 5.6 Control inputs of all agents

Table 5.1 The times of controller updates

Inputs	Times
u_1	518
u_2	555
u_3	621
u_4	623
u_5	620

5.5 Conclusion

This chapter establishes distributed event-triggered control protocols for group consensus to deal with uncertainties of multi-agent systems. Event-triggered controller without continuous updates and communication is designed to achieve the consensus and restrain the uncertainties, respectively. In addition, the Zeno behavior can be excluded with the proposed controller.

References

Dimarogonas, D. V., Frazzoli, E., and Johansson, K. H. (2012). Distributed event-triggered control for multi-agent systems, *IEEE Transactions on Automatic Control***57**, 5, pp. 1291–1297.

Du, J. H., Xu, L., Liu, Y. R., Song, Y., and Fan, X. L. (2016). Linear optimal filtering for time-delay networked systems subject to missing measurements with individual occurrence probability, *Neurocomputing***214**, pp. 767–774.

Fan, Y. and Yang, J. Y. (2016). Average consensus of multi-agent systems with self-triggered controllers, *Neurocomputing***177**, pp. 33–39.

Fan, Y., Feng, G., Wang, Y., and Song, C. (2013). Distributed event-triggered control of multi-agent systems with combinational measurements, *Automatica***49**, 2, pp. 671–675.

Garcia, E., Cao, Y. C., and Casbeer, D. W. (2016). Decentralised event-triggered consensus of double integrator multi-agent systems with packet losses and communication delays, *IET Control Theory and Applications***10**, 15, pp. 1835–1843.

Grune, L. and Muller, F. (2009). An algorithm for event-based optimal feedback control, in *Proceedings of the 48th IEEE Conference on Decision and Control, 2009 held jointly with the 2009 28th Chinese Control Conference*, Proceedings of the 48th IEEE Conference on Decision and Control, 2009 held jointly with the 2009 28th Chinese Control Conference, pp. 5311–5316, decision and Control, 2009 held jointly with the 2009 28th Chinese Control Conference. CDC/CCC 2009. Proceedings of the 48th IEEE Conference on.

Heemels, W., Sandee, J. H., and Van den Bosch, P. (2008). Analysis of event-driven controllers for linear systems, *International Journal of Control***81**, 4, pp. 571–590.

Hu, W. F. and Liu, L. (2017). Cooperative output regulation of heterogeneous linear multi-agent systems by event-triggered control, *IEEE Transactions on Cybernetics***47**, 1, pp. 105–116.

Hu, J., Wang, Z. D., Liu, S., and Gao, H. J. (2016b). A variance-constrained approach to recursive state estimation for time-varying complex networks with missing measurements, *Automatica***64**, pp. 155–162.

Hu, W. F., Liu, L., and Feng, G. (2017). Output consensus of heterogeneous linear multi-agent systems by distributed event-triggered/self-triggered strategy, *IEEE Transactions on Cybernetics***47**, 8, pp. 1914–1924.

Li, Z. K., Duan, Z. S., and Lewis, F. L. (2014). Distributed robust consensus control of multi-agent systems with heterogeneous matching uncertainties, *Automatica***50**, 3, pp. 883–889.

Li, H. Q., Chen, G., Huang, T. W., Zhu, W., and Xiao, L. (2016b). Event-triggered consensus in nonlinear multi-agent systems with nonlinear dynamics and directed network topology, *Neurocomputing***185**, pp. 105–112.

Li, H. Q., Chen, G., Dong, Z. Y., and Xia, D. W. (2016a). Consensus analysis of multiagent systems with second-order nonlinear dynamics and general directed topology: An event-triggered scheme, *Information Sciences***370**, pp. 598–622.

Liu, J., Yu, Y., Wang, Q., and Sun, C. (2017a). Fixed-time event-triggered consensus control for multi-agent systems with nonlinear uncertainties, *Neurocomputing***260**, pp. 497–504.

Liu, S., Quevedo, D. E., and Xie, L. H. (2017c). Event-triggered distributed constrained consensus, *International Journal of Robust and Nonlinear Control***27**, 16, pp. 3043–3060.

Liuzza, D., Dimarogonas, D. V., di Bernardo, M., and Johansson, K. H. (2016). Distributed model based event-triggered control for synchronization of multi-agent systems, *Automatica***73**, pp. 1–7.

Mazo, M. and Tabuada, P. (2011). Decentralized event-triggered control over wireless sensor/actuator networks, *IEEE Transactions on Automatic Control***56**, 10, pp. 2456–2461.

Meng, X. Y. and Chen, T. W. (2013). Event based agreement protocols for multi-agent networks, *Automatica***49**, 7, pp. 2125–2132.

Mu, C. X., Ni, Z., Sun, C. Y., and He, H. B. (2017a). Data-driven tracking control with adaptive dynamic programming for a class of continuous-time nonlinear systems, *IEEE Transactions on Cybernetics***47**, 6, pp. 1460–1470.

Mu, C. X., Wang, D., and He, H. B. (2017b). Novel iterative neural dynamic programming for data-based approximate optimal control design, *Automatica***81**, pp. 240–252.

Ren, W. and Beard, R. W. (2005). Consensus seeking in multiagent systems under dynamically changing interaction topologies, *IEEE Transactions on Automatic Control***50**, 5, pp. 655–661.

Tabuada, P. (2007). Event-triggered real-time scheduling of stabilizing control tasks, *IEEE Transactions on Automatic Control***52**, 9, pp. 1680–1685.

Wu, Y. Q., Meng, X. Y., Xie, L. H., Lu, R. Q., Su, H. Y., and Wu, Z. G. (2017). An input-based triggering approach to leader-following problems, *Automatica***75**, pp. 221–228.

Xie, G., Liu, H., Wang, L., and Jia, Y. (2009). Consensus in networked multi-agent systems via sampled control: Fixed topology case, in *American Control Conference*, American Control Conference (Institute of Electrical and Electronics Engineers Inc., St. Louis, MO, United states), pp. 3902–3907.

Yang, D. P., Ren, W., Liu, X. D., and Chen, W. S. (2016). Decentralized event-triggered consensus for linear multi-agent systems under general directed graphs, *Automatica***69**, pp. 242–249.

Yu, W. W., Chen, G. R., and Cao, M. (2011). Consensus in directed networks of agents with nonlinear dynamics, *IEEE Transactions on Automatic Control***56**, 6, pp. 1436–1441.

Zhang, X. M. and Han, Q. L. (2014). Event-triggered dynamic output feedback control for networked control systems, *IET Control Theory and Applications***8**, 4, pp. 226–234.

Zhang, Z. Q., Hao, F., Zhang, L., and Wang, L. (2014). Consensus of linear multi-agent systems via event-triggered control, *International Journal of Control***87**, 6, pp. 1243–1251.

Zhu, W. and Jiang, Z. P. (2015). Event-based leader-following consensus of multi-agent systems with input time delay, *IEEE Transactions on Automatic Control***60**, 5, pp. 1362–1367.

Chapter 6
Distributed Convex Nonsmooth Optimization for Multi-agent System Based on Proximal Operator

Recently, distributed optimization for multi-agent systems has been an interesting topic and attracted more and more attention due to its wide range of applications such as smart grids, sensor networks and mobile manipulators (see Chen and Kai 2018; Fang et al. 2018). In distributed optimization, the objective is to optimize the global cost function composed of the sum of all local cost functions, each of which is only known by its own local agent.

As a typical concern in multi-agent systems, the distributed optimization has been used to describe many practical collective tasks. One of its representative formulations is to make the consensus state be the optimal point of a sum of individual convex functions. Such problem is also called optimal consensus sometimes, and many discrete-time dynamics algorithms have been proposed especially since 2010 (see Nedic and Ozdaglar 2009; Lou et al. 2014 for distributed subgradient methods, Yuan et al. 2011, 2016 for distributed primal-dual subgradient methods, and Shi et al. (2014) for distributed alternating direction method of multipliers).

In addition to discrete-time optimization algorithms, growing interest can be witnessed in the study of continuous-time distributed optimization (see Shi et al. 2013; Kia et al. 2015; Liu and Wang 2015; Qiu et al. 2016; Liu et al. 2017). On one hand, the design and analysis of continuous-time algorithms are often more convenient in comparison with discrete-time ones due to the well-developed stability theory. On the other hand, the discrete-time algorithms can be viewed as the discretization of continuous-time algorithms if appropriate sample intervals are selected. Additionally, some practical systems are continuous-time, in which the optimization algorithms need to be continuous. In particular, Shi et al. (2013) solved the problem of computing the intersection of convex sets by a distributed continuous-time optimization algorithm, while Kia et al. (2015) proposed a distributed continuous-time gradient algorithm for unconstrained optimization problems. Moreover, Liu and Wang (2015)-Qiu et al. (2016) gave several projected distributed algorithms for various constrained optimization problems, and Liu et al. (2017) developed a collective neurodynamic method for constrained distributed optimization.

Q. Wang et al., *Distributed Cooperative Control and Optimization for Multi-Agent Systems*, https://doi.org/10.1007/978-981-96-0950-5_6

In practice, many interesting problems can be formulated as convex optimization problems of the form: $\min_{x \in R^n} \sum_{i=1}^{n} f_i(x_i) + g_i(x_i)$, in which f_i, $(i = 1, \ldots, n)$ are differentiable convex functions defined from $f : R^n \rightarrow R$ while $g_i, i = 1, \ldots, n$ are non-differentiable convex functions defined from $g : R^n \rightarrow R$. It is meaningful and challenging to develop distributed algorithms to solve the non-differentiable convex optimization problems.

There are a lot of advantages to develop proximal algorithms (see Parikh and Stephen 2014; Boyd et al. 2010). Firstly, proximal algorithms can deal with the cases where the functions are nonsmooth; secondly, the simple proximal operators can be chosen so they are easy to understand, derive and implement for optimization problems; thirdly, they are amenable to solve large scale problems, such as big data and large-dimensional problems.

The objective of this chapter is to propose Lipschitz continuous algorithms for multi-agent systems to solve non-differentiable convex optimization problems. In previous work (Zeng et al. 2017), a novel distributed continuous-time projected algorithm is developed to study the distributed optimization problem of a sum of nonsmooth convex cost functions with local constraints. In Zeng et al. (2018), two novel continuous-time distributed subgradient-based algorithms with projected output feedback and derivative feedback, respectively, to solve the extended monotropic optimization problem. However, the methods proposed in Zeng et al. (2017, 2018) are both nonsmooth and double-integrator multi-agent systems are not considered. In this chapter, both singe- and double-integrator dynamics for multi-agent systems are considered to achieve the optimal consensus. The main contributions of this chapter are listed as the following: Firstly, we design a Lipschitz continuous algorithm based on proximal operator and derivative feedback for the single-integral multi-agent systems, Secondly, based on the first-order proximal operator, the optimization algorithm for double-integral multi-agent systems is developed, which provides a new idea for non-differentiable optimization problems. Thirdly, the convergence and correctness of the proposed algorithm are proved with Lyapunov stability theory.

6.1 Mathematical Preliminaries

6.1.1 *Convex Analysis*

A set $K \subseteq R^n$ is convex if

$$ax + (1 - a)y \in K, \quad x, y \in K, \quad 0 \leq a \leq 1.$$

A function $f(\cdot) : K \rightarrow R$, where K is a convex set, is said to be convex if for any $0 \leq a \leq 1$

$$f(ax + (1-a)y) \leq af(x) + (1-a)f(y), \quad \forall x, y \in K.$$

A differentiable function f is convex over K if

$$f(x) - f(y) \geq \nabla f(y)^T (x - y), \forall x, y \in K,$$

and f is strictly convex if the above inequality is strict whenever $x \neq y$, and f is ω-strongly convex ($\omega > 0$) if

$$(\nabla f(x) - \nabla f(y))^T (x - y) \geq \omega \|x - y\|^2, \quad \forall x, y \in K.$$

A function $f : R^n \to R$ is Lipschitz with constant $\theta > 0$, or simply θ-Lipschitz over a set $K \in R^n$ if

$$|f(x) - f(y)| \leq \theta \|x - y\|, \quad \forall x, y \in K.$$

6.1.2 Graph Theory

For a graph $\mathcal{G}$ with N vertices, denote the graph $\mathcal{G} = \{\mathcal{V}, \mathcal{E}\}$. $\mathcal{V} = \{v_1, v_2, \ldots, v_n\}$ represents the set of nodes, and $\mathcal{E} = \{(v_i, v_j)|v_i, v_j \in \mathcal{V}\}$ is the set of edges. $\mathcal{G}$ is called undirected if $(v_i, v_j) \in \mathcal{E} \Leftrightarrow (v_j, v_i) \in \mathcal{E}$. Meanwhile, if there is an edge between v_i and v_j, then it is said that node v_i and node v_j are adjacent. The adjacency matrix $\mathcal{A} = [a_{ij}] \in R^{n \times n}$ associated with $\mathcal{G}$ is defined as $a_{ij} = 1$ if $(v_j, v_i) \in \mathcal{E}$. For each agent v_i, N_i is the cardinality of its neighbor set and $N_{min} = \min\{N_i\}$. The Laplacian matrix $L = [l_{ij}] \in R^{n \times n}$ is defined as

$$\begin{cases} l_{ii} = \sum\limits_{j=1, j \neq i}^{n} a_{ij}, & i = j, \\ l_{ij} = -a_{ij}, & i \neq j. \end{cases}$$

Obviously, for an undirected graph, L is symmetric and positive semi-definite, i.e. $L^T = L \geq 0$. The row sum of L is zero; thus, the corresponding eigenvector of eigenvalue $\lambda_1(\mathcal{G}) = 0$ is $\mathbf{1}$,i.e., $L\mathbf{1} = \mathbf{0}$. For a connected graph, there exists exactly one zero eigenvalue for L, and the eigenvalues are sorted in ascending order $0 = \lambda_1 < \lambda_2 \leq \cdots \leq \lambda_N$.

6.1.3 Notations

For simplicity, R denotes the set of real numbers; R^m denotes the set of m-dimensional real column vectors; $R^{n \times m}$ denotes the set of $n \times m$ real matrices; 0 is used to denote zero matrices of appropriate size with zero vectors and zero numbers as special cases; I_n stands for an identity matrix with appropriate dimension; $(\cdot)^T$ denotes the transpose of matrix $\cdot$; and $A \otimes B$ refers to the Kronecker product of matrices A and B, respectively. Furthermore, $\|\cdot\|$ represents the Euclidean norm.

Assume f is a lower semi-continuous convex function and the proximal operator associated with the convex function f is defined by

$$\text{prox}_f(v) = \arg\min_x (f(x) + 1/2\|x - v\|^2).$$

Define the indicator function of a closed convex set Ω as $I_\Omega(x) = 0$ if $x \in \Omega$ and $I_\Omega(x) = +\infty$ otherwise. We have $\text{prox}_{I_\Omega}(v) = P_\Omega(v)$, where $P_\Omega(v) = \arg\min_{x\in\Omega} \|x - v\|$ is the projection operator. Let $\partial f(x)$ denote the subgradient of $f(\cdot)$ at x. Therefore, $\partial f(x)$ is monotone, implying $(\partial f(x) - \partial f(y))^T(x - y) \geq 0$ for all x, y. $x = \text{prox}_f(v)$ is equivalent to $v - x \in \partial f(x)$.

6.2 First-Order Proximal-Gradient Flow

In this section, we first consider a general group of N agents with first-order dynamics by the following

$$\dot{x}_i(t) = u_i(t), \tag{6.1}$$

where $i = 1, \ldots, N$, $x_i \in R^n$ and $u_i \in R^n$ respectively denote the states and control inputs of the i-th agent. Each agent i is associated with a private convex cost function $f_i(x_i) : R^n \to R$.

The objective of the distributed optimization problem is to design the control input u_i for every agent, so that the sum of $f_i + g_i$ can be minimized via cooperation. The problem is equivalent to the following problem on R^{Nn}:

$$\begin{aligned} &\min \sum_{i=1}^{N} f_i(x_i) + g_i(x_i), \\ &\text{subject to} \quad Lx = 0_{Nn}, \end{aligned} \tag{6.2}$$

where $f : R^n \to R$ is a twice differentiable convex function and $g : R^n \to R$ is a non-differentiable function.

Assumption 6.1 Graph $\mathcal{G}$ is undirected and connected.

Assumption 6.2 For each agent i, the function f_i is differentiable and m-strongly convex satisfying $(a - b)^T(\nabla f_i(a) - \nabla f_i(b)) \geq m\|a - b\|^2, \forall a, b \in R^n$.

Assumption 6.3 For each agent i, there exists a constant $\theta_i > 0$, such that $\|\nabla f_i(a) - \nabla f_i(b)\| \leq \theta_i \|a - b\|, \forall a, b \in R^n$.

Now consider the following algorithm

$$\begin{cases} u_i = \text{prox}_{g_i}[x_i - \alpha\nabla f_i(x_i) - w_i \\ \qquad - \beta\sum_{j=1}^{N} a_{ij}(x_i - x_j)] - x_i, \\ \dot{w}_i = \alpha\beta\sum_{j=1}^{N} a_{ij}(x_i - x_j + \dot{x}_i - \dot{x}_j) \\ \sum_{i=1}^{N} w_i(0) = 0, \end{cases} \tag{6.3}$$

where w_i is introduced to eliminate the consensus error caused by the gradient differences.

Define $x = [x_1^T, \ldots, x_N^T]^T$,$y = [y_1^T, \ldots, y_N^T]$ and $w = [w_1^T, \ldots, w_N^T]$. Then the closed-loop system for system (6.1) is written as

$$\begin{cases} \dot{x} = \text{prox}_g[x - \alpha\nabla\tilde{f}(x) - \beta(L \otimes I_n)x - w] - x, \\ \dot{w} = \alpha\beta(L \otimes I_n)(x + \dot{x}), \end{cases} \tag{6.4}$$

where $\tilde{f}(x) = \sum_{j=1}^{N} f_i(x_i)$.

Assume that (x^*, w^*) is the equilibrium point of the closed-loop system (6.4) and it has that

$$\begin{cases} Lx^* = 0, \\ \text{prox}_g[x^* - \alpha\nabla\tilde{f}(x^*) - w^*] = x^*. \end{cases} \tag{6.5}$$

Define the following coordinate transformation

$$\tilde{x} = x - x^*,$$
$$\tilde{w} = w - w^*.$$

Thus, we have

$$\begin{cases} \dot{\tilde{x}} = prox_g[\tilde{x} - \alpha h - \tilde{w} - \beta(L \otimes I_n)\tilde{x}] - \tilde{x}, \\ \dot{\tilde{w}} = \alpha\beta(L \otimes I_n)(\tilde{x} + \dot{\tilde{x}}), \end{cases} \tag{6.6}$$

where $h = \nabla\tilde{f}(x) - \nabla\tilde{f}(1_n \otimes x^*)$.

If system (6.6) is asymptotically stable, then it has that the agent states $x_i, i = 1, \ldots, N$ will converge to the optimization point x^*. Thus, the following is to prove the stability of the system (6.6).

Define an orthogonal matrix Q satisfying $Q^T Q = I_n$, and with the orthogonal transformations it is derived

$$\hat{x} = (Q^T \otimes I_n)\tilde{x},$$
$$\hat{w} = (Q^T \otimes I_n)\tilde{w}.$$

With the above transformation, (6.6) can be rewritten as

$$\begin{cases} \dot{\hat{x}} = \text{prox}_g[\hat{x} - \alpha(Q^T \otimes I_n)h - \hat{w} \\ \qquad - \beta(Q^T L Q \otimes I_n)\hat{x}] - \hat{x}, \\ \dot{\hat{w}} = \alpha\beta(Q^T L Q \otimes I_n)(\hat{x} + \dot{\hat{x}}). \end{cases} \tag{6.7}$$

Theorem 6.1 *Suppose Assumptions 6.1–6.3 hold. When the parameters* α, β, θ *and* m *satisfy the following inequality*

$$\alpha m + \beta\lambda_2 - \theta^2\alpha^2/4 > 0,$$

the distributed optimization problem (6.2) for the first-order multi-agent system (6.1) can be solved with the distributed control protocol (6.3) for any initial conditions $x_i(0), \dot{x}_i(0), w_i(0) \in R^n$ *satisfying* $\sum_{i=1}^{N} w_i(0) = 0$*, which implies*

$$\lim_{t\to\infty} x_i(t) = x^*, \ \lim_{t\to\infty} \dot{x}_i(t) = 0,$$

for all $i \in \mathcal{V}$*, where* $x^* = \arg\min_{x\in R^n} f(x)$.

Proof Choose the following Lyapunov candidate function

$V = \frac{1}{2}\hat{x}^T\hat{x} + \frac{1}{2\alpha\beta}\hat{w}^T(Q^T L Q)^{-1}\hat{w}$

It results from (6.4) and (6.5) that

$$[x - \alpha\nabla\tilde{f}(x) - w - \beta Lx] - x - \dot{x} \in \partial_g(x + \dot{x}), \tag{6.8}$$

$$[x^* - \alpha\nabla\tilde{f}(x^*) - w^*] - x^* \in \partial_g(x^*). \tag{6.9}$$

Hence, from (6.8) and (6.9) and the convexity of $\partial_g(x)$, it has

$$\begin{aligned} &[-\alpha(\nabla\tilde{f}(x) - \nabla\tilde{f}(x^*)) - (w - w^*) \\ &-\beta Lx - \dot{x}]^T(x + \dot{x} - x^*) \geq 0. \end{aligned} \tag{6.10}$$

Further,

$$\begin{aligned}
&-\alpha(\nabla\tilde{f}(x)-\nabla\tilde{f}(x^*))^T(x-x^*)-(w-w^*)^T(x-x^*)\\
&-\beta x^T L(x-x^*)-\dot{y}^T(x-x^*)\geq\\
&\alpha(\nabla\tilde{f}(x)-\nabla\tilde{f}(x^*))^T\dot{y}+(w-w^*)^T\dot{y}\\
&+\beta x^T L\dot{y}+\dot{y}^T\dot{y}. \qquad (6.11)
\end{aligned}$$

Therefore, combining the system (6.7), the derivative of V is described as

$$\begin{aligned}
\dot{V}=&\hat{x}^T\dot{x}+1/(\alpha\beta)\hat{w}^T(Q^T L Q)^{-1}\dot{\hat{w}}\\
=&\hat{x}^T\dot{x}+\hat{w}^T\hat{x}+\hat{w}^T\dot{\hat{x}}\\
\leqslant&-\alpha(\nabla\tilde{f}(x)-\nabla\tilde{f}(x^*))^T(x-x^*)-\beta\hat{x}^T L\hat{x}\\
&-\alpha(\nabla\tilde{f}(x)-\nabla\tilde{f}(x^*))^T\hat{x}-\beta\hat{x}^T L\dot{x}\\
&-\alpha(\nabla\tilde{f}(x)-\nabla\tilde{f}(x^*))^T\dot{x}-\dot{x}^T\dot{x}. \qquad (6.12)
\end{aligned}$$

From the Assumptions, it is known that

$$(x-x^*)^T(\nabla\tilde{f}(x)-\nabla\tilde{f}(x^*))\geqslant m(x-x^*)^T(x-x^*),$$

$$\|\nabla\tilde{f}(x)-\nabla\tilde{f}(x^*)\|\leqslant\theta\|x-x^*\|,$$

and

$$\begin{aligned}
&-\alpha(\nabla\tilde{f}(x)-\nabla\tilde{f}(x^*))^T\dot{x}\\
&\leqslant\theta^2\alpha^2/4\hat{x}^T\hat{x}+\dot{x}^T\dot{x}. \qquad (6.13)
\end{aligned}$$

Therefore, it is easy to obtain

$$\begin{aligned}
\dot{V}\leqslant&-(\alpha m+\beta\lambda_2-\theta^2\alpha^2/4)\hat{x}^T\hat{x}\\
\leq&0. \qquad (6.14)
\end{aligned}$$

Thus, $\hat{x}$ and $\hat{w}$ can converge to zero asymptotically, which implies the system (6.1) can converge to the equilibrium point with the proposed algorithm (6.3) and the optimization problem (6.2) can be solved.

6.3 Second-Order Proximal-Gradient Flow

In this section, a general group of N agents with second-order dynamics are considered:

$$\ddot{x}_i(t) = u_i(t), \tag{6.15}$$

where $i = 1, \ldots, N$, $x_i \in R^n$ and $u_i \in R^n$ respectively denote the states and control inputs of the i-th agent. Each agent i is associated with a private convex cost function $f_i(x_i) : R^n \to R$. The goal of the distributed optimization problem is described in (6.2)

Now consider the following algorithm

$$\begin{cases} \dot{x}_i = y_i, \\ \dot{y}_i = u_i, \\ u_i = \text{prox}_{g_i}[x_i - \alpha \nabla f_i(x_i) - w_i \\ \quad - \beta \sum\limits_{j=1}^{N} a_{ij}(x_i - x_j) - k y_i] - x_i, \\ \dot{w}_i = \alpha\beta \sum\limits_{j=1}^{N} a_{ij}(x_i - x_j + \dot{y}_i - \dot{y}_j), \\ \sum\limits_{i=1}^{N} w_i(0) = 0, \end{cases} \tag{6.16}$$

where w_i is introduced to eliminate the consensus error caused by the gradient differences.

Define $x = [x_1^T, \ldots, x_N^T]^T$,$y = [y_1^T, \ldots, y_N^T]$ and $w = [w_1^T, \ldots, w_N^T]$. Then the closed-loop system for system (6.15) is written as

$$\begin{cases} \dot{x} = y, \\ \dot{y} = \text{prox}_g[x - \alpha \nabla \tilde{f}(x) - \beta(L \otimes I_n)x - w - ky] - x, \\ \dot{w} = \alpha\beta(L \otimes I_n)(x + \dot{y}), \end{cases} \tag{6.17}$$

where $\tilde{f}(x) = \sum\limits_{j=1}^{N} f_i(x_i)$.

Assume that (x^*, y^*, w^*) is the equilibrium point of the closed-loop system (6.17) and it can have that

$$\begin{cases} y^* = 0, \\ Lx^* = 0, \\ \text{prox}_g[x^* - \alpha \nabla \tilde{f}(x^*) - w^*] = x^*. \end{cases} \tag{6.18}$$

Define the following coordinate transformation

$$\begin{cases} \tilde{x} = x - x^*, \\ \tilde{y} = y - y^*, \\ \tilde{w} = w - w^*. \end{cases}$$

Thus, we have

$$\begin{cases} \dot{\tilde{x}} = \tilde{y}, \\ \dot{\tilde{y}} = \text{prox}_g[\tilde{x} - k\tilde{y} - \alpha h - \tilde{w} - \beta(L \otimes I_n)\tilde{x}] - \tilde{x}, \\ \dot{\tilde{w}} = \alpha\beta(L \otimes I_n)(\tilde{x} + \dot{\tilde{y}}), \end{cases} \tag{6.19}$$

where $h = \nabla \tilde{f}(x) - \nabla \tilde{f}(1_n \otimes x^*)$.

If the above system (6.19) is asymptotically stable, then we can have that the agent states $x_i, i = 1, \ldots, N$ will converge to the optimisation point x^*. Thus, the following is to prove the stability of the system (6.19).

Define an orthogonal matrix Q satisfying $Q^T Q = I_n$, and with the orthogonal transformations it is derived

$$\hat{x} = (Q^T \otimes I_n)\tilde{x},$$
$$\hat{y} = (Q^T \otimes I_n)\tilde{y},$$
$$\hat{w} = (Q^T \otimes I_n)\tilde{w}.$$

With the above transformation, the (6.19) can be rewritten as

$$\begin{cases} \dot{\hat{x}} =\hat{y}, \\ \dot{\hat{y}} =\text{prox}_g[\hat{x} - k\hat{y} - \alpha(Q^T \otimes I_n)h - \hat{w} \\ \qquad - \beta(Q^T L Q \otimes I_n)\hat{x}] - \hat{x}, \\ \dot{\hat{w}} =\alpha\beta(Q^T L Q \otimes I_n)(\hat{x} + \dot{\hat{y}}). \end{cases} \tag{6.20}$$

Theorem 6.2 *Suppose Assumptions 6.1–6.3 hold. When the parameters* $0 < \alpha < \sqrt{2}$, β, θ *and* m *satisfy the following inequality*

$$\alpha m + \beta\lambda_2 - 1 - \theta^2 - \beta^2\lambda_N^2 > 0,$$

the distributed optimization problem (6.2) for the second-order multi-agent system (6.15) is solved with the distributed control protocol (6.16) for any initial conditions $x_i(0), \dot{x}_i(0), w_i(0) \in R^n$ *satisfying* $\sum_{i=1}^{N} w_i(0) = 0$, *which implies*

$$\lim_{t\to\infty} x_i(t) = x^*, \lim_{t\to\infty} \dot{x}_i(t) = 0,$$

for all $i \in \mathcal{V}$, *where* $x^* = arg\min_{x \in R^n} f(x)$.

Proof Choose the following Lyapunov candidate function
$V = \frac{k}{2}\hat{x}^T\hat{x} + \frac{k}{2}\hat{y}^T\hat{y} + \frac{1}{2\alpha\beta}\hat{w}^T(Q^TLQ)^{-1}\hat{w}$.
By (6.17) and (6.18), it has

$$[x - \alpha\nabla\tilde{f}(x) - w - \beta Lx - ky] - x - \dot{y} \in \partial_g(x + \dot{y}), \tag{6.21}$$

$$[x^* - \alpha\nabla\tilde{f}(x^*) - w^*] - x^* \in \partial_g(x^*). \tag{6.22}$$

Hence, from (6.21) and (6.22) and the convexity of $\partial_g(x)$, it is derived that

$$\begin{aligned}&[-\alpha(\nabla\tilde{f}(x) - \nabla\tilde{f}(x^*)) - (w - w^*)\\&-\beta Lx - ky - \dot{y}]^T(x + \dot{y} - x^*) \geq 0.\end{aligned} \tag{6.23}$$

Furthermore,

$$\begin{aligned}&-\alpha(\nabla\tilde{f}(x) - \nabla\tilde{f}(x^*))^T(x - x^*) - (w - w^*)^T(x - x^*)\\&-\beta x^TL(x - x^*) - ky^T(x - x^*) - \dot{y}^T(x - x^*) \geq\\&\alpha(\nabla\tilde{f}(x) - \nabla\tilde{f}(x^*))^T\dot{y} + (w - w^*)^T\dot{y}\\&+\beta x^TL\dot{y} + ky^T\dot{y} + \dot{y}^T\dot{y}.\end{aligned} \tag{6.24}$$

Therefore, the derivative of V along the system (6.20) can be described as:

$$\begin{aligned}\dot{V} =&k\hat{x}^Ty + k\hat{y}^T\dot{y} + 1/(\alpha\beta)\hat{w}^T(Q^TLQ)^{-1}\dot{\hat{w}}\\=&k\hat{x}^Ty + k\hat{y}^T\dot{y} + \hat{w}^T\hat{x} + \hat{w}^T\dot{\hat{y}}\\\leqslant&-\alpha(\nabla\tilde{f}(x) - \nabla\tilde{f}(x^*))^T(x - x^*) - \beta\hat{x}^TL\hat{x}\\&-\hat{x}^T\dot{y} - \beta\hat{x}^TL\dot{y} - \alpha(\nabla\tilde{f}(x) - \nabla\tilde{f}(x^*))^T\dot{y} - \dot{y}^T\dot{y}.\end{aligned} \tag{6.25}$$

It is known that

$$\alpha(\nabla\tilde{f}(x) - \nabla\tilde{f}(x^*))^T(x - x^*) \geqslant m(x - x^*)^T(x - x^*),$$

$$\|\nabla\tilde{f}(x) - \nabla\tilde{f}(x^*)\| \leqslant \theta\|x - x^*\|,$$

and

$$-\hat{x}^T\dot{y} - \beta\hat{x}^TL\dot{y} - \alpha(\nabla\tilde{f}(x) - \nabla\tilde{f}(x^*))^T\dot{y}$$

$$\leqslant \hat{x}^T\hat{x} + 1/4\dot{y}^T\dot{y} + \theta^2\hat{x}^T\hat{x} + \alpha^2/4\dot{y}^T\dot{y}$$
$$+ \beta^2\lambda_N^2\hat{x}^T\hat{x} + 1/4\dot{y}^T\dot{y}. \quad (6.26)$$

Therefore, it has

$$\begin{aligned}\dot{V} \leqslant &- (\alpha m + \beta\lambda_2 - 1 - \theta^2 - \beta^2\lambda_N^2)\hat{x}^T\hat{x} \\ &- (1/2 - 1/4\alpha^2)\dot{y}^T\dot{y} \\ &\leq 0. \end{aligned} \quad (6.27)$$

Thus, $\hat{x}$, $\hat{y}$ and $\hat{w}$ can converge to zero asymptotically, which implies the system (6.15) can converge to the equilibrium point with the proposed algorithm (6.16) and the optimization problem (6.2) can be solved.

6.4 Numerical Simulation

In this section, we give a numerical example to illustrate the given theoretical results and show the performance of the distributed proximal continuous-time control proposed.

Consider a connected and undirected multi-agent system grouped by five agents which is shown as Fig. 6.1. The local cost functions are given by

$$\begin{aligned} g_i(x) =&|x|, (i = 1, \ldots, 5), \\ f_1(x) =&(x-4)^2, \\ f_2(x) =&(x-2)^2 + x, \\ f_3(x) =&x^2 + 3, \\ f_4(x) =&(x-1)^2, \\ f_5(x) =&2(x+1)^2 - x. \end{aligned}$$

Obviously, the selected $g_i(x)(i = 1, \ldots, 5)$ is non-differentiable and the gradient of the given $f_i(x)(i = 1, \ldots, 5)$, above satisfy the Lipschitz condition. The Laplacian matrix of the communication network graph is described by

$$L = \begin{pmatrix} 2 & -1 & -1 & 0 & 0 \\ -1 & 2 & 0 & 0 & -1 \\ -1 & 0 & 2 & -1 & 0 \\ 0 & 0 & -1 & 2 & -1 \\ 0 & -1 & 0 & -1 & 2 \end{pmatrix}.$$

Case 1: Considering the algorithm (6.3) for the first-order multi-agent system (6.1), select the parameters $\alpha = 1, \beta = 1$, arbitrary initial values $w_i(0)$ satisfying

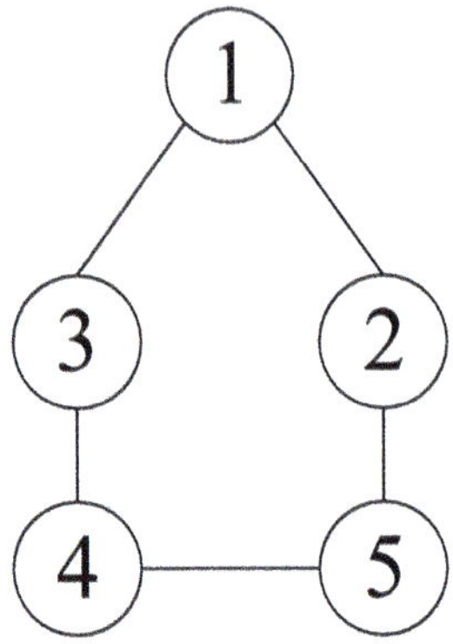

Fig. 6.1 The communication topology

$\sum_{i=1}^{5} w_i(0) = 0, i = 1, \ldots, 5$ and arbitrary initial states $x_i(0)$. Figures 6.2 and 6.3 show the trajectories of the generalized coordinates and the gradients of these five agents under control protocol (6.3), respectively. Through the simulations, we can see that the states of the agents will converge to $x^* = 1.2500$ as well as the global optimal objective function $\min \sum_{i=1}^{5} f_i(x) + g_i(x) = 29.1250$.

Case 2: Considering the algorithm (6.16) for the second-order multi-agent system (6.15), select the parameters $\alpha = 1, \beta = 1, k = 4$, arbitrary initial values $w_i(0)$ satisfying $\sum_{i=1}^{5} w_i(0) = 0, i = 1, \ldots, 5$ and arbitrary initial states $x_i(0)$. Figures 6.4, 6.5 and 6.6 show the trajectories of the generalized coordinates, their derivations and the gradients of these five agents under control protocol (6.16), respectively. Through

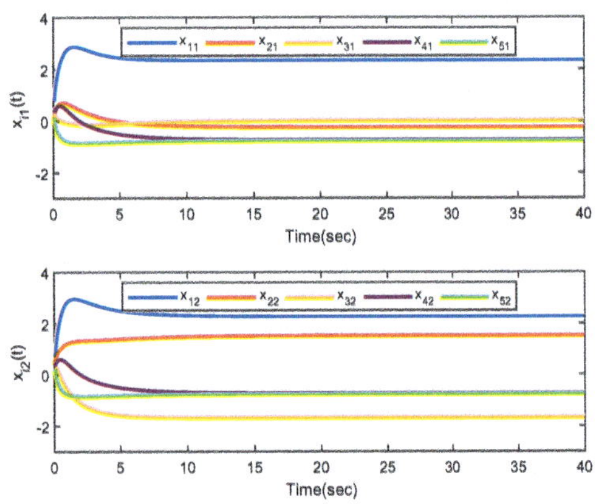

Fig. 6.2 The trajectories of the states

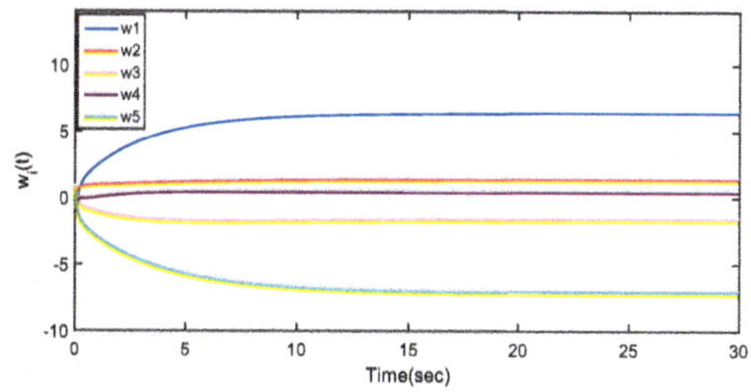

Fig. 6.3 The trajectories of the gradients

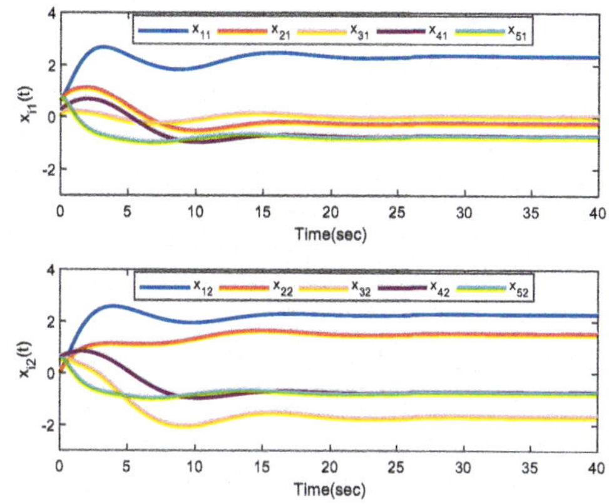

Fig. 6.4 The trajectories of the states

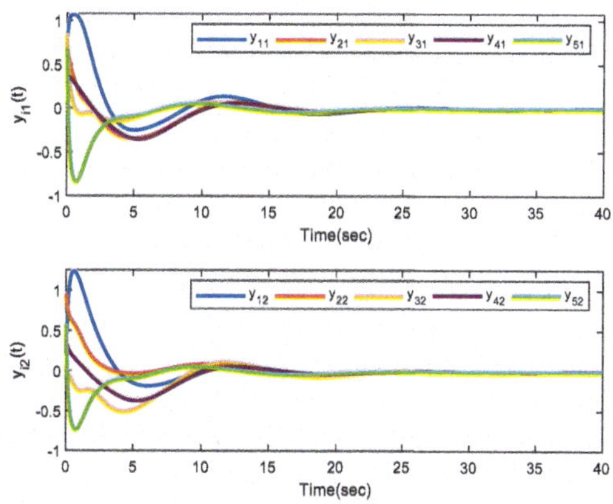

Fig. 6.5 The trajectories of the states' derivations

the simulations, we can see that the states of the agents will converge to $x^* = 1.2500$ as well as the global optimal objective function $\min \sum_{i=1}^{5} f_i(x) + g_i(x) = 29.1250$.

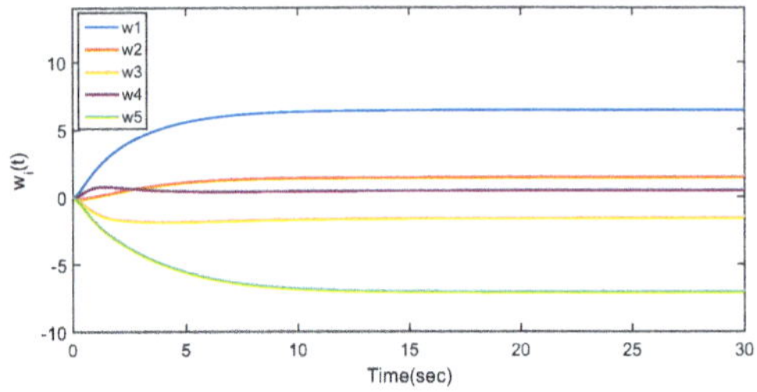

Fig. 6.6 The trajectories of the gradients

6.5 Conclusion

In this chapter, the distributed optimization problem for both single-integrator and double-integrator multi-agent systems are studied. Different from most existing results, the distributed optimization control methods are based on the proximal operator and derivative feedback, which can solve the nonsmooth convex optimization. In the future, the algorithm based on the proximal operator will be studied for high-order multi-agent systems with nonlinear uncertainties or switching topologies.

References

Boyd, S., Parikh, N., Chu, E., Peleato, B., and Eckstein, J. (2010). "Distributed optimization and statistical learning via the alternating direction method of multipliers," *Foundations and Trends in Machine Learning***3**, 1, pp. 1–122.

Chen, J. and Kai, S. X. (2018). "Cooperative transportation control of multiple mobile manipulators through distributed optimization," *Science China-Information Sciences***61**, 12.

Fang, H., Shang, C. S., and Chen, J. (2018). "An optimization-based shared control framework with applications in multi-robot systems," *Science China-Information Sciences***61**, 1.

Kia, S. S., Cortes, J., and Martinez, S. (2015). "Distributed convex optimization via continuous-time coordination algorithms with discrete-time communication," *Automatica***55**, pp. 254–264.

Liu, Q. S. and Wang, J. (2015). "A second-order multi-agent network for bound-constrained distributed optimization," *IEEE Transactions on Automatic Control***60**, 12, pp. 3310–3315.

Liu, Q. S., Yang, S. F., and Wang, J. (2017). "A collective neurodynamic approach to distributed constrained optimization," *IEEE Transactions on Neural Networks and Learning Systems***28**, 8, pp. 1747–1758.

Lou, Y. C., Shi, G. D., Johansson, K. H., and Hong, Y. G. (2014). "Approximate projected consensus for convex intersection computation: Convergence analysis and critical error angle," *IEEE Transactions on Automatic Control***59**, 7, pp. 1722–1736.

Nedic, A. and Ozdaglar, A. (2009). "Distributed subgradient methods for multi-agent optimization," *IEEE Transactions on Automatic Control***54**, 1, pp. 48–61.

Parikh, N. and Stephen, B. (2014). "Proximal algorithms," *Foundations and Trends extregistered in Optimization***3**, 1, pp. 127–239.

Qiu, Z. R., Liu, S., and Xie, L. H. (2016). "Distributed constrained optimal consensus of multi-agent systems," *Automatica***68**, pp. 209–215.

Shi, W., Ling, Q., Yuan, K., Wu, G. and Yin, W. (2014). "On the linear convergence of the admm in decentralized consensus optimization," *IEEE Transactions on Signal Processing***62**, 7, pp. 1750–1761.

Shi, G. D., Johansson, K. H., and Hong, Y. G. (2013). "Reaching an optimal consensus: Dynamical systems that compute intersections of convex sets," *IEEE Transactions on Automatic Control* **58**, 3, pp. 610–622.

Yuan, D. M., Xu, S. Y., and Zhao, H. Y. (2011). "Distributed primal-dual subgradient method for multiagent optimization via consensus algorithms," *IEEE Transactions on Systems Man and Cybernetics Part B-Cybernetics* **41**, 6, pp. 1715–1724.

Yuan, D. M., Ho, D., and Xu, S. Y. (2016). "Regularized primal-dual subgradient method for distributed constrained optimization," *IEEE Transactions on Cybernetics* **46**, 9, pp. 2109–2118.

Zeng, X. L., Yi, P., and Hong, Y. G. (2017). "Distributed continuous-time algorithm for constrained convex optimizations via nonsmooth analysis approach," *IEEE Transactions on Automatic Control* **62**, 10, pp. 5227–5233.

Zeng, X. L., Yi, P., Hong, Y. G., and Xie, L. H. (2018). "Distributed continuous-time algorithms for nonsmooth extended monotropic optimization problems," *SIAM Journal on Control and Optimization* **56**, 6, pp. 3973–3993.

Chapter 7
Distributed Proximal-Gradient Algorithms for Nonsmooth Convex Optimization of Second-Order Multi-agent Systems

Nonsmooth optimization (NSO) refers to the general problem of minimizing (or maximizing) functions that are typically not differentiable. Due to the complexity of the real world, functions involved in practical applications are often nonsmooth. For example, a wide range of problems in the fields of machine learning and multi-agent system can fall into NSO form Tibshirani et al. (2005), Lin et al. (2017). Therefore, NSO problems for multi-agent systems have attracted more and more attention due to the wide applications. The subgradient method is an effective way to solve NSO problems (Shor 2012; Nedic et al. 2010; Burger et al. 2012; Ram et al. 2010; Carlone et al. 2014). The basic idea of the subgradient method is to generalize smooth methods by replacing the gradient with an arbitrary subgradient. A distributed subgradient algorithm based on the projection method was proposed for distributed optimization with nonsmooth objective functions and local bound constraints (Nedic et al. 2010). A distributed simplex algorithm was proposed for solving degenerate linear programs and multiagent assignments (Burger et al. 2012). The stochastic subgradient projection algorithms were presented to solve the distributed optimization problem with the presence of noise (Ram et al. 2010). A distributed algorithm was proposed for random convex programming with a large number of randomly extracted constraints (Carlone et al. 2014). These aforementioned results are involved in discrete-time methods.

The proximal method is applicable and especially well suited to problems involving big data and large-dimensional problems (Boyd et al. 2010; Parikh and Stephen 2014). It is a generalized form of projection, which provides a way to solve nondifferentiable convex optimization problems with continuous-time algorithms. It is known that continuous-time algorithms are much more popular, especially when the optimization problem may be achieved with continuous-time physical systems. On one hand, the design and analysis of continuous-time algorithms are often easier than discrete-time ones due to the well-development of stability theory. On the other hand, the discrete versions of continuous-time algorithms are also useful as long as a proper step size is chosen. Thus, continuous-time algorithms are often viewed

Q. Wang et al., *Distributed Cooperative Control and Optimization for Multi-Agent Systems*, https://doi.org/10.1007/978-981-96-0950-5_7

as prototypes of discrete-time algorithms. In particular, Shi et al. (2013) solved the convex intersection computation problem by a distributed continuous-time optimization algorithm based on projection method. A continuous-time multiagent network on weight-balanced digraphs was presented for distributed unconstrained convex optimization (Gharesifard and Cortes 2014). A one-layer projection neural network was presented to solve nonsmooth optimization problems with linear equalities and bound constraints (Liu and Wang 2013). Based on either projection or differentiated projection, two classes of continuous-time algorithms were proposed to solve this distributed optimization problem in an initialization-free and scalable manner (Yi et al. 2016).

In practice, many optimization problems for multi-agent systems are involved in various constraints, such as optimal consensus and resource allocation, which can be viewed as distributed optimization with equality constraints. The continuous-time methods have been developed for distributed nonsmooth optimization with constraints (Yi et al. 2015; Li et al. 2019; Zeng et al. 2018; Liang et al. 2018). A novel distributed continuous-time algorithm was proposed by using the Karush-Kuhn-Tucker (KKT) condition combined with the Lagrangian multiplier method (Yi et al. 2015), but the local objective functions and local constraints are assumed to be twice continuously differentiable, which belongs to smooth optimization. For nonsmooth optimization, a distributed continuous-time algorithm was designed using subgradient projections for a general nonsmooth monotropic optimization problem (Li et al. 2019). The distributed projection algorithms based on derivative feedback were proposed for optimization with set constraints and equality constraints (Zeng et al. 2018). The projected primal-dual algorithms were proposed in the form of differential inclusions to solve the distributed nonsmooth constrained optimization problems (Liang et al. 2018). The results mentioned above (Li et al. 2019; Zeng et al. 2018; Liang et al. 2018) are based on the subgradient of the nonsmooth cost function, which is discontinuous and may cause violent vibration in the systems. Besides, the results above are not involved in second-order multi-agent systems.

The objective of this paper is to propose smooth algorithms for distributed nonsmooth convex optimization problems of multi-agent systems. To avoid the subgradient, the nonsmooth convex objective function is divided into two parts: a twice continuously differentiable function and a nonsmooth convex function. Based on the proximal operator and Lagrangian function, the distributed proximal-gradient algorithms are proposed to solve the consensus and resource allocation problems, respectively. Moreover, the convergence analysis of the proposed algorithms is carried out based on Lyapunov stability theory. The main contributions of the paper are three folds:

(i) Distributed smooth proximal-gradient algorithms based on proximal operator and Lagrangian function are proposed for nonsmooth convex optimization of second-order multi-agent systems;

(ii) The convergence and correctness of the proposed algorithm are proved based on Lyapunov stability theory. A novel view of primal-dual type algorithm is provided, which can avoid nonsmooth analysis;

(iii) The proposed algorithms with proximal operator in this paper are more general than the projection algorithms (Yi et al. 2015; Li et al. 2019; Zeng et al. 2018). Projection algorithms can be viewed as the special form of the proximal-gradient algorithms.

The remainder of the paper is organized as follows: In Sect. 7.1, the notation, some basic mathematical definitions, and theoretical results are presented. In Sect. 7.2, the optimal consensus problem based on proximal-gradient algorithm is solved for second-order multi-agent systems. In Sect. 7.3, resource allocation problems of multi-agent systems are solved with the proposed smooth proximal-gradient algorithms. In Sect. 7.4, the proposed algorithms are applied to distributed optimal consensus and resource allocation problems, respectively. Numerical simulations are shown to demonstrate the effectiveness of the proposed algorithms. Finally, Sect. 7.5 concludes the paper.

7.1 Mathematical Preliminaries

7.1.1 Notation

Specifically, $\mathbb{R}$ denotes the set of real numbers; $\mathbb{R}^n$ denotes the set of n-dimensional real column vectors; $\mathbb{R}^{n\times m}$ denotes the set of n-by-m real matrices; I_n denotes the $n \times n$ identity matrix; $(\cdot)^{\mathrm{T}}$ denotes transpose. We write rank A for the rank of the matrix A, range(A) for the range of the matrix A, ker(A) for the kernel of the matrix A, $\mathbf{1}_n$ for the ones vector with dimension $n \times 1$, $\mathbf{0}_n$ for the $n \times 1$ zeros vector, and $A \otimes B$ for the Kronecker product of matrices A and B. Furthermore, $\|\cdot\|$ denotes the Euclidean norm; $\|\cdot\|_p$ denotes the p-norm where $p \geq 1$; $A > 0\,(A \geq 0)$ implies that matrix $A \in \mathbb{R}^{n\times n}$ is positive definite (positive semi-definite); $\overline{\mathcal{S}}$ denotes the closure of the subset $\mathcal{S} \subset \mathbb{R}^n$; int$(\mathcal{S})$ denotes the interior of the subset $\mathcal{S}$; dim$(\mathcal{S})$ denotes the dimension of the vector space $\mathcal{S}$; dist$(p, \mathcal{M})$ denotes the distance from a point p to the set $\mathcal{M}$, that is, $\mathrm{dist}(p, \mathcal{M}) \triangleq \inf_{x\in\mathcal{M}} \|p - x\|$; $x(t) \to \mathcal{M}$ as $t \to \infty$ implies that $x(t)$ approaches the set $\mathcal{M}$, that is, for each $\epsilon > 0$ there exists $T > 0$ such that $\mathrm{dist}(x(t), \mathcal{M}) < \epsilon$ for all $t > T$.

7.1.2 Graph Theory

A weighted undirected graph $\mathcal{G}$ is denoted by $\mathcal{G}(\mathcal{V}, \mathcal{E}, A)$, where $\mathcal{V} = \{1, 2, \ldots, n\}$ is a set of nodes, $\mathcal{E} \subset \mathcal{V} \times \mathcal{V}$ is a set of edges, $A = [a_{i,j}] \in \mathbb{R}^{n\times n}$ is a weighted adjacency matrix such that $a_{i,j} = a_{j,i} > 0$ if $(j, i) \in \mathcal{E}$ and $a_{i,j} = 0$ otherwise. The Laplacian matrix is $L = D - A$, where $D \in \mathbb{R}^{n\times n}$ is diagonal with (i, i)-th element $D_{i,i} = \sum_{j=1}^{n} a_{i,j}$, $i \in \{1, 2, \ldots, n\}$. If the weighted graph $\mathcal{G}$ is undirected and connected, then $L = L^T \geq 0$, $\text{rank}(L) = n - 1$, λ_{max} denotes the largest eigenvalue of the Laplacian matrix L.

7.1.3 Convex Analysis

Convexity plays an important role in the study of optimization problems. We introduce convex definitions and Lipschitz continuity of functions as follows.

Definition 7.1 (*Convex function*) Ozdaglar (2003), Zhang and Hong (2015) The function $f(\cdot) : S \to \mathbb{R}$ is called convex if for any $0 \leq \mu \leq 1$

$$f(\mu x + (1 - \mu)y) \leq \mu f(x) + (1 - \mu) f(y), \quad \forall x, y \in S,$$

is satisfied, where S is a convex set.

If a differentiable function $f(\cdot)$ is convex over S, it has

$$f(x) - f(y) \geq \nabla f(y)^T (x - y), \quad \forall x, y \in S,$$

and $f(\cdot)$ is strictly convex if

$$f(x) - f(y) > \nabla f(y)^T (x - y),$$

whenever $x \neq y$. Furthermore, $f(\cdot)$ is said to be m-strongly convex ($m > 0$) if

$$(\nabla f(x) - \nabla f(y))^T (x - y) \geq m\|x - y\|^2, \quad \forall x, y \in S,$$

is satisfied, where $\nabla f(x)$ is the gradient of $f(\cdot)$ at x.

Definition 7.2 (*θ-Lipschitz*) Wei et al. (2020), Hui et al. (2009) A function $f(\cdot) : S \to \mathbb{R}^n$ is Lipschitz with constant $\theta > 0$, or simply θ-Lipschitz over the set S if

$$\|f(x) - f(y)\| \leq \theta\|x - y\|, \quad \forall x, y \in S,$$

is satisfied.

7.1.4 Proximal Operator

Let $g(\cdot)$ be a lower semi-continuous convex function. The proximal operator of $g(\cdot)$ is Parikh and Stephen (2014)

$$\text{prox}_g(v) = \arg\min_x \{g(x) + \frac{1}{2}\|x - v\|^2\}.$$

Define the indicator function of a closed convex set Ω as $I_\Omega(x) = 0$ if $x \in \Omega$ and $I_\Omega(x) = +\infty$ otherwise. We have $\text{prox}_{I_\Omega}(v) = P_\Omega(v)$, where $P_\Omega(v) = \arg\min_{x\in\Omega} \|x - v\|$ is the projection operator.

Let $\partial g(x)$ denote the subdifferential of $g(\cdot)$ at x. Then $\partial g(x)$ is monotone, that is,

$$(p_x - p_y)^T(x - y) \geq 0$$

for all x, y, where $p_x \in \partial g(x)$, and $p_y \in \partial g(y)$. $x = \text{prox}_g(v)$ is equivalent to

$$v - x \in \partial g(x). \tag{7.1}$$

7.1.5 Convergence Property

Consider a system

$$\dot{x}(t) = \phi(x(t)), \quad x(0) = x_0, \quad t \geq 0, \tag{7.2}$$

where $\phi : \mathbb{R}^q \to \mathbb{R}^q$ is Lipschitz continuous.

The following result is a version of the invariance principle (Hui et al. 2009).

Lemma 7.1 *Let $\mathcal{D}$ be a compact, positive invariant set with respect to system (7.2), $V : \mathbb{R}^q \to \mathbb{R}$ be a continuously differentiable function, and $x(\cdot)$ be a solution of (7.2) with $x(0) = x_0 \in \mathcal{D}$. Assume $\dot{V}(x) \leq 0$, $\forall x \in \mathcal{D}$. Define $\mathcal{Z} = \{x \in \mathcal{D} : \dot{V}(x) = 0\}$, and $\mathcal{M}$ is the largest invariant subset of $\overline{\mathcal{Z}} \bigcap \mathcal{D}$, where $\overline{\mathcal{Z}}$ is the closure of $\mathcal{Z} \subset \mathbb{R}^q$. If $\dot{V}(x) \leq 0$ for all $x \in \mathcal{D}$, then* $\text{dist}(x(t), \mathcal{M}) \to 0$ *as $t \to \infty$.*

7.2 Proximal-Gradient Algorithm for Distributed Optimal Consensus

7.2.1 Problem Formulation

Optimal consensus is an essential problem of distributed optimization, which means all agents reach consensus at a state that minimizes the sum of the objective functions of all agents. In this section, distributed optimal consensus with nonsmooth objective is investigated for a class of second-order multi-agent systems.

Consider a group of n agents over a graph $\mathcal{G}$. Each agent is described by second-order integrator as

$$\ddot{x}_i = u_i, \quad i \in \{1, 2, \ldots, n\}$$

where $x_i \in \mathbb{R}^q$ is the state of i-th agent. Define $x \triangleq [x_1^T, x_2^T, \ldots, x_n^T]^T \in \mathbb{R}^{nq}$ and $L_q \triangleq L \otimes I_q \in \mathbb{R}^{nq\times nq}$, where $L \in \mathbb{R}^{n\times n}$ is the Laplacian matrix of $\mathcal{G}$. The optimal consensus problem is described as follows

$$\min_{x\in\mathbb{R}^{nq}} (f(x) + g(x)), \tag{7.3a}$$

$$s.t. \quad L_q x = \mathbf{0}_{nq}. \tag{7.3b}$$

where $f(x) = \sum_{i=1}^{n} f_i(x_i)$ and $g(x) = \sum_{i=1}^{n} g_i(x_i)$. $f_i(\cdot) : \mathbb{R}^q \to \mathbb{R}$ is a twice differentiable convex function and $g_i(\cdot) : \mathbb{R}^q \to \mathbb{R}$ is a nonsmooth convex function. The constraint presented in (7.3b) indicates that all the states reach consensus. Each agent i only knows its own local information x_i and exchanges information with its neighbors in distributed manner.

Remark 7.1 The problem described by (7.3) above has vast applications. If $g_i(x_i)$ is a regularized function such as $g_i(x_i) = \mu\|x_i\|_1$, the problem becomes a very common model in machine learning such as l_1 regularized loss minimization and LASSO (Tibshirani et al. 2005).

Assumption 7.1 To ensure the wellposedness of the problem and algorithm, it is assumed that

(1) $f_i(\cdot)$ is a twice differentiable convex function and $\nabla f_i(\cdot)$ is θ_i-Lipschitz for all $i \in \{1, 2, \ldots, n\}$;
(2) $f_i(\cdot)$ is m_i-strongly convex;
(3) $g_i(\cdot)$ is (nonsmooth) lower semi-continuous convex for all $i \in \{1, 2, \ldots, n\}$, whose proximal operator is assumed to be easy to obtain;
(4) $\mathcal{G}$ is connected and undirected;
(5) There exists at least one finite solution to problem (7.3).

7.2.2 Distributed Continuous-Time Algorithm Based on Proximal Operator and Gradient

The distributed algorithm is proposed as

$$\begin{cases} \dot{x}_i = y_i, \\ \dot{y}_i = \text{prox}_{g_i}(x_i - \nabla f_i(x_i) - \sum_{j=1}^{n} a_{i,j}(\eta_i - \eta_j) - c\sum_{j=1}^{n} a_{i,j}(x_i - x_j) - \beta y_i) - x_i - \gamma y_i, \\ \dot{\eta}_i = \sum_{j=1}^{n} a_{i,j}(x_i - x_j) + \gamma \sum_{j=1}^{n} a_{i,j}(y_i - y_j), \end{cases} \tag{7.4}$$

where $y_i \in \mathbb{R}^q$, $\eta_i \in \mathbb{R}^q$, and $a_{i,j}$ is the (i, j)-th element of the weighted adjacency matrix of graph $\mathcal{G}$. Define $y \triangleq [y_1^T, y_2^T, \ldots, y_n^T]^T \in \mathbb{R}^{nq}$, $\eta \triangleq [\eta_1^T, \eta_2^T, \ldots, \eta_n^T]^T \in \mathbb{R}^{nq}$, $m \triangleq \min\{m_1, m_2, \cdots, m_n\}$ and $\theta \triangleq \max\{\theta_1, \theta_2, \cdots, \theta_n\}$.

The compact form of algorithm (7.13) is

$$\begin{cases} \dot{x} = y, \\ \dot{y} = \mathrm{prox}_g(x - \nabla f(x) - L_q\eta - cL_qx - \beta y) - x - \gamma y, \\ \dot{\eta} = L_q(x + \gamma y). \end{cases} \tag{7.5}$$

Remark 7.2 Algorithm (7.5) is a primal-dual proximal gradient method to solve the saddle point of the Lagrangian function $\mathcal{L}(x, \eta) = f(x) + g(x) + \eta^T L_q x$, where x is the primal variable and η is the dual variable. Note that a saddle point satisfies that $0 \in \partial_x \mathcal{L}(x, \eta)$ and $\nabla_\eta \mathcal{L}(x, \eta) = 0$.

Remark 7.3 Since the proximal operator $\mathrm{prox}_g(\cdot)$ is continuous and nonexpansive, the proposed algorithm is a Lipschitz continuous system even though the problem (7.3) is a nonsmooth problem.

Then, we arrive at the following lemma by the Karush-Kuhn-Tucker (KKT) condition of convex optimization problems (Ruszczyski 2006).

Lemma 7.2 *Suppose Assumption 7.1 holds. A feasible point $x^* \in \mathbb{R}^{nq}$ is a minimizer to problem (7.3) if and only if there exist x^*, y^*, η^* such that (x^*, y^*, η^*) is an equilibrium point of algorithm* (7.5), *where $y^* = \mathbf{0}_{nq}$ and $\eta^* \in \mathbb{R}^{nq}$.*

Proof Necessity: if (x^*, y^*, η^*) is an equilibrium point of algorithm (7.5), according to the property of proximal operator in (7.1) and algorithm (7.5) it yields

$$\begin{cases} -\nabla f(x^*) - L_q\eta^* - cL_qx^* \in \partial g(x^*), \\ L_qx^* = \mathbf{0}_{nq}, \\ y^* = \mathbf{0}_{nq}. \end{cases}$$

Furthermore, it has $\mathbf{0}_{nq} \in \nabla f(x^*) + \partial g(x^*) + L_q\eta^*$. According to the Karush-Kuhn-Tucker (KKT) condition of convex optimization problems, x^* is a minimizer to problem (7.3).

Sufficiency: if x^* is a solution of the problem (7.3), according to the Karush-Kuhn-Tucker (KKT) condition, there exists $\eta^* \in \mathbb{R}^{nq}$ such that

$$\begin{cases} \mathbf{0}_{nq} \in \nabla f(x^*) + L_q\eta^* + \partial g(x^*), \\ L_qx^* = \mathbf{0}_{nq}. \end{cases}$$

With the property of proximal operator in (7.1), it is derived that

$$\begin{cases} \mathrm{prox}_g(x^* - \nabla f(x^*) - L_q\eta^*) = x^*, \\ L_qx^* = \mathbf{0}_{nq}, \end{cases} \tag{7.6}$$

Combining the fact that $cL_q x^* = \mathbf{0}_{nq}$ and $y^* = \mathbf{0}_{nq}$, the above (7.6) is equal to

$$\begin{cases} y^* = \mathbf{0}_{nq}, \\ \text{prox}_g(x^* - \nabla f(x^*) - L_q\eta^* - cL_q x^* - \beta y^*) - x^* - \gamma y^* = \mathbf{0}_{nq}, \\ L_q x^* = \mathbf{0}_{nq}, \end{cases}$$

which means that (x^*, y^*, η^*) is an equilibrium point of algorithm (7.5). □

Define a function

$$V(x, y, \eta) = V_1(x, y) + V_2(x, y, \eta) \tag{7.7}$$

with

$$\begin{aligned} V_1(x, y) = & \frac{\beta + 2\gamma}{2}\|y\|^2 + \frac{\beta + \gamma}{2}\|x - x^*\|^2 + \frac{c\gamma}{2}x^T L_q x \\ & + \gamma[f(x) - f(x^*) - (x - x^*)^T \nabla f(x^*)], \\ V_2(x, y, \eta) = & \frac{1}{2}\|\eta - \eta^*\|^2 + y^T L_q(\eta - \eta^*) + y^T(x - x^* + cL_q x), \end{aligned}$$

where (x^*, y^*, η^*) is an equilibrium point of algorithm (7.5). Then the following Lemma is derived.

Lemma 7.3 *Suppose Assumption 7.1 holds. The function $V(x, y, \eta)$ along the trajectory of algorithm* (7.5) *satisfies that*

$$\begin{aligned} \dot{V}(x, y, \eta) \le & -(\nabla f(x) - \nabla f(x^*))^T(\dot{y} + x - x^*) - (\gamma(\beta + \gamma) - 1)\|y\|^2 - \|\dot{y}\|^2 \\ & - x^T(cL_q - 0.5L_q^2)x + y^T(cL_q + (0.5 + \gamma)L_q^2)y. \end{aligned} \tag{7.8}$$

Proof Suppose (x^*, y^*, η^*) is an equilibrium point of algorithm (7.5). Then

$$\begin{cases} y^* = \mathbf{0}_{nq}, \\ \text{prox}_g(x^* - \nabla f(x^*) - L_q\eta^*) = x^*, \\ L_q x^* = \mathbf{0}_{nq}. \end{cases} \tag{7.9}$$

It follows from (7.1), (7.5), and (7.9) that

$$\begin{aligned} (x - \nabla f(x) - L_q\eta - cL_q x - \beta y) - x - \dot{y} - \gamma y \in \partial g(x + \dot{y} + \gamma y), \\ (x^* - \nabla f(x^*) - L_q\eta^*) - x^* \in \partial g(x^*). \end{aligned}$$

Recall that $g(\cdot)$ is a convex function. $\partial g(\cdot)$ is monotone, and hence,

$$(-(\nabla f(x) - \nabla f(x^*)) - L_q(\eta - \eta^*) - cL_q x - (\beta + \gamma)y - \dot{y})^T(x + \dot{y} + \gamma y - x^*) \ge 0.$$

Furthermore, it is derived

$$\begin{aligned}&\gamma(\nabla f(x)-\nabla f(x^*))^T y+(\beta+\gamma)y^T(x-x^*)+c\gamma y^T L_q x+(\beta+2\gamma)y^T\dot{y}\\&\le -(\nabla f(x)-\nabla f(x^*))^T\dot{y}-c\dot{y}^T L_q x-\gamma(\beta+\gamma)y^T y-\|\dot{y}\|^2-(\dot{y}+\gamma y)^T L_q(\eta-\eta^*)\\&\quad-(\nabla f(x)-\nabla f(x^*))^T(x-x^*)-(x-x^*)^T L_q(\eta-\eta^*)-cx^T L_q x-\dot{y}^T(x-x^*).\end{aligned}$$

Note that $L_q x^*=\mathbf{0}_{nq}$, $y^T\dot{y}=\frac{1}{2}\frac{\mathrm{d}}{\mathrm{dt}}\|y\|^2$, $y^T(x-x^*)=\frac{1}{2}\frac{\mathrm{d}}{\mathrm{dt}}\|x-x^*\|^2$, $y^T L_q x=\frac{1}{2}\frac{\mathrm{d}}{\mathrm{dt}}x^T L_q x$, and $(\nabla f(x)-\nabla f(x^*))^T y=\frac{\mathrm{d}}{\mathrm{dt}}[f(x)-f(x^*)-(x-x^*)^T\nabla f(x^*)]$. Hence,

$$\begin{aligned}\dot{V}_1(x,y)&=(\beta+2\gamma)y^T\dot{y}+(\beta+\gamma)y^T(x-x^*)+c\gamma y^T L_q x+\gamma(\nabla f(x)-\nabla f(x^*))^T y\\&\le -(\nabla f(x)-\nabla f(x^*))^T(x-x^*)-\gamma(\beta+\gamma)\|y\|^2-\|\dot{y}\|^2-cx^T L_q x\\&\quad-(x-x^*)^T L_q(\eta-\eta^*)-\dot{y}^T(x-x^*)\\&\quad-(\nabla f(x)-\nabla f(x^*))^T\dot{y}-(\dot{y}+\gamma y)^T L_q(\eta-\eta^*)-c\dot{y}^T L_q x\\&=-(\nabla f(x)-\nabla f(x^*))^T(\dot{y}+x-x^*)-\gamma(\beta+\gamma)\|y\|^2-\|\dot{y}\|^2-cx^T L_q x\\&\quad-(x-x^*+\dot{y}+\gamma y)^T L_q(\eta-\eta^*)-\dot{y}^T(x-x^*+cL_q x)\\&=-(\nabla f(x)-\nabla f(x^*))^T(\dot{y}+x-x^*)-\gamma(\beta+\gamma)\|y\|^2-\|\dot{y}\|^2-cx^T L_q x\\&\quad-(x+\dot{y}+\gamma y)^T L_q(\eta-\eta^*)-\dot{y}^T(x-x^*+cL_q x).\end{aligned}$$

It follows from (7.5) that

$$\begin{aligned}\dot{V}_2(x,y,\eta)&=(\eta-\eta^*)^T L_q(x+\gamma y)\\&\quad+\dot{y}^T L_q(\eta-\eta^*)+y^T L_q^2(x+\gamma y)+\dot{y}^T(x-x^*+cL_q x)+\|y\|^2+y^T L_q y\\&=(\eta-\eta^*)^T L_q(x+\dot{y}+\gamma y)+y^T L_q^2(x+\gamma y)\\&\quad+\dot{y}^T(x-x^*+cL_q x)+\|y\|^2+cy^T L_q y.\end{aligned}$$

Combining the fact that $y^T L_q^2 x\le 0.5y^T L_q^2 y+0.5x^T L_q^2 x$, the result (7.8) is obtained by summing $\dot{V}_1(x,y)$ and $\dot{V}_2(x,y,\eta)$ as follows:

$$\begin{aligned}\dot{V}(x,y,\eta)&=\dot{V}_1(x,y)+\dot{V}_2(x,y,\eta)\\&\le -(\nabla f(x)-\nabla f(x^*))^T(\dot{y}+x-x^*)-[\gamma(\beta+\gamma)-1]\|y\|^2 \qquad (7.10)\\&\quad-\|\dot{y}\|^2-cx^T L_q x+y^T L_q^2 x+\gamma y^T L_q^2 y+cy^T L_q y\\&\le -(\nabla f(x)-\nabla f(x^*))^T(\dot{y}+x-x^*)-[\gamma(\beta+\gamma)-1]\|y\|^2-\|\dot{y}\|^2-cx^T L_q x\\&\quad+0.5y^T L_q^2 y+0.5x^T L_q^2 x+\gamma y^T L_q^2 y+cy^T L_q y\\&\le -(\nabla f(x)-\nabla f(x^*))^T(\dot{y}+x-x^*)-[\gamma(\beta+\gamma)-1]\|y\|^2-\|\dot{y}\|^2\\&\quad-x^T(cL_q-0.5L_q^2)x+y^T(cL_q+(0.5+\gamma)L_q^2)y. \qquad (7.11)\end{aligned}$$

Then we are ready to present the following main result:

Theorem 7.1 *Suppose Assumption 7.1 holds and consider algorithm* (7.5). *If positive constants m, θ, β, γ and c satisfy the following inequalities:*

$$
\begin{cases}
\gamma \geq 1, \\
c \geq 0.5\lambda_{\max}, \\
m - \dfrac{\theta^2}{4} > 0, \\
\beta + 2\gamma - 1 - c\lambda_{\max} - 2\lambda_{\max}^2 \geq 0, \\
\gamma(\beta + \gamma) - 1 - c\lambda_{\max} - (0.5 + \gamma)\lambda_{\max}^2 \geq 0,
\end{cases} \tag{7.12}
$$

it has:

(i) The function $V(x, y, \eta)$ is positive definite, and $V(x, y, \eta) = 0$ if and only if $(x, y, \eta) = (x^*, y^*, \eta^*)$;

(ii) The system achieves distributed optimal consensus.

Remark 7.4 The parameters β, γ and c are easy to select if the communication topology among agents is given. No matter how large $\lambda_{\max}$ is, we can always adjust parameter β adaptively to satisfy the last two inequalities in (7.12) after γ and c are chosen. θ and m are related to the given objective function, which can be adjusted through scaling transformation. For example, the minimization of objective function $f(x) + g(x)$ equals to the minimization of $\rho(f(x) + g(x))$ for any $\rho > 0$. Hence, θ and m can be scaled by ρ to satisfy the inequality constraint $\tilde{m} - \dfrac{\tilde{\theta}^2}{4} > 0$, where $\tilde{m} = \rho m$ and $\tilde{\theta} = \rho\theta$. Accordingly, the algorithm (7.13) is modified as

$$
\begin{cases}
\dot{x}_i = y_i, \\
\dot{y}_i = \mathrm{prox}_{\tilde{g}_i}(x_i - \nabla \tilde{f}_i(x_i) - \sum_{j=1}^{n} a_{i,j}(\eta_i - \eta_j) - c\sum_{j=1}^{n} a_{i,j}(x_i - x_j) - \beta y_i) - x_i - \gamma y_i, \\
\dot{\eta}_i = \sum_{j=1}^{n} a_{i,j}(x_i - x_j) + \gamma \sum_{j=1}^{n} a_{i,j}(y_i - y_j),
\end{cases} \tag{7.13}
$$

where $\tilde{f}_i = \rho f_i$ and $\tilde{g}_i = \rho g_i$.

Proof (i) Let $V(x, y, \eta)$ be defined as (7.7). Since $f(x)$ is a convex function, it follows that

$$
f(x) - f(x^*) - (x - x^*)^T \nabla f(x^*) \geq 0. \tag{7.14}
$$

Therefore,

$$
\begin{aligned}
V_1(x, y) = {} & \frac{\beta + 2\gamma}{2}\|y\|^2 + \frac{\beta + \gamma}{2}\|x - x^*\|^2 + \frac{c\gamma}{2}x^T L_q x \\
& + \gamma[f(x) - f(x^*) - (x - x^*)^T \nabla f(x^*)] \\
\geq {} & 0.
\end{aligned}
$$

With the facts that $x^T y \geq -x^T x - \frac{1}{4}y^T y$ and $x^T y \geq -\frac{1}{2}x^T x - \frac{1}{2}y^T y$, it is derived that

$$
\begin{aligned}
V_2(x, y, \eta) &= \frac{1}{2}\|\eta - \eta^*\|^2 + y^T L_q(\eta - \eta^*) + y^T(x - x^* + cL_q x), \\
&\geq \frac{1}{2}\|\eta - \eta^*\|^2 - y^T L_q^2 y - \frac{1}{4}\|\eta - \eta^*\|^2 \\
&- \frac{1}{2}y^T y - \frac{1}{2}\|x - x^*\|^2 - \frac{c}{2}y^T L_q y - \frac{c}{2}x^T L_q x.
\end{aligned}
$$

Combining $V_1(x, y)$ and $V_2(x, y, \eta)$, it follows that

$$
\begin{aligned}
V(x, y, \eta) \geq &\left(\frac{\beta + 2\gamma - 2\lambda_{\max}^2 - 1 - c\lambda_{\max}}{2}\right)\|y\|^2 + \frac{c\gamma - c}{2}x^T L_q x \\
&+ \frac{\beta + \gamma - 1}{2}\|x - x^*\|^2 + \tfrac{1}{4}\|\eta - \eta^*\|^2.
\end{aligned}
$$

In view of the conditions $\gamma > 1$ and $\beta + 2\gamma - 1 - c\lambda_{\max} - 2\lambda_{\max}^2 > 0$, it follows that $V(x, y, \eta)$ is positive definite. Clearly, $V(x, y, \eta) = 0$ if and only if $(x, y, \eta) = (x^*, y^*, \eta^*)$.

(ii) It follows from Lemma 7.3 that

$$
\begin{aligned}
\dot{V}(x, y, \eta) &\leq -(\nabla f(x) - \nabla f(x^*))^T(\dot{y} + x - x^*) - (\gamma(\beta + \gamma) - 1)\|y\|^2 \\
&- \|\dot{y}\|^2 - x^T(cL_q - 0.5L_q^2)x + y^T(cL_q + (0.5 + \gamma)L_q^2)y \\
&\leq -m(x - x^*)^T(x - x^*) - (\nabla f(x) - \nabla f(x^*))^T\dot{y} \\
&- (\gamma(\beta + \gamma) - 1)\|y\|^2 - \|\dot{y}\|^2 \\
&- x^T(cL_q - 0.5L_q^2)x + y^T(cL_q + (0.5 + \gamma)L_q^2)y \\
&\leq -m(x - x^*)^T(x - x^*) + \frac{\theta^2}{4}(x - x^*)^T(x - x^*) + \dot{y}^T\dot{y} \\
&- [\gamma(\beta + \gamma) - 1]y^2 - \|\dot{y}\|^2 \\
&- x^T(cL_q - 0.5L_q^2)x + (c\lambda_{\max} + (0.5 + \gamma)\lambda_{\max}^2)y^T y \\
&\leq -\varepsilon(x - x^*)^T(x - x^*) - x^T(cL_q - 0.5L_q^2)x - \varpi y^T y \qquad (7.15)
\end{aligned}
$$

where $\varepsilon = m - \dfrac{\theta^2}{4}$ and $\varpi = \gamma(\beta + \gamma) - 1 - c\lambda_{\max} - (0.5 + \gamma)\lambda_{\max}^2$.

When $c \geq 0.5\lambda_{\max}$, it follows that $-x^T(cL_q - 0.5L_q^2)x \leq 0$. Furthermore, with the given conditions $m - \dfrac{\theta^2}{4} > 0$ and $\gamma(\beta + \gamma) - 1 - c\lambda_{\max} - (0.5 + \gamma)\lambda_{\max}^2 \geq 0$ in (7.12), it has that $\dot{V}(x, y, \eta) \leq 0$.

For all $(x, y, \eta) \in \mathbb{R}^{nq} \times \mathbb{R}^{nq} \times \mathbb{R}^{nq}$, $V(x, y, \eta)$ is positive definite. $V(x, y, \eta) = 0$ if and only if $(x, y, \eta) = (x^*, y^*, \eta^*)$, and $V(x, y, \eta) \to \infty$ as $(x, y, \eta) \to \infty$. Hence, the solution (x, y, η) is bounded for all $t > 0$. Let $\mathcal{Z} = \{(x, y, \eta) \in \mathbb{R}^{nq} \times \mathbb{R}^{nq} \times \mathbb{R}^{nq} : \dot{V}(x, y, \eta) = 0\} \subset \{(x, y, \eta) \in \mathbb{R}^{nq} \times \mathbb{R}^{nq} \times \mathbb{R}^{nq} : \varepsilon(x - x^*)^T(x - x^*) = 0,\ x^T(cL_q - 0.5L_q^2)x = 0,\ \varpi y^T y = 0\}$. Let $\mathcal{M}$ be the largest invariant subset of $\overline{\mathcal{Z}}$. It follows from Lemma 7.1 that $(x(t), y(t), \eta(t)) \to \mathcal{M}$ as $t \to \infty$. Note that $\varepsilon(x - x^*)^T(x - x^*) > 0$ if $x \neq x^*$ because $\varepsilon > 0$. Hence, $x(t) \to x^*$ as $t \to \infty$, and the distributed optimal consensus for problem (7.3) is solved.

7.3 Proximal-Gradient Algorithm for Resource Allocation Problem

7.3.1 *Problem Formulation*

Resource allocation problem is an important problem in smart grids. In smart grids, there are n generators supplying power to satisfy the demands of users, and each generator has its own local cost function due to the different efficiency of generators. It can be modeled as distributed optimization with equality constraints for multi-agent systems, shown as:

$$\min_{x_i \in \mathbb{R}^q} \sum_{i=1}^{n} (f_i(x_i) + g_i(x_i)), \tag{7.16a}$$

$$s.t. \sum_{i=1}^{n} w_i x_i = d_0, \tag{7.16b}$$

where $x_i \in \mathbb{R}^q$, $w_i \in \mathbb{R}^{p \times q}$, $\sum_{i=1}^{n} d_i = d_0 \in \mathbb{R}^p$, $f_i(\cdot) : \mathbb{R}^q \to \mathbb{R}$ is a twice differentiable convex function and $g_i(\cdot) : \mathbb{R}^q \to \mathbb{R}$ is a nonsmooth convex function.

Remark 7.5 When the w_i is an identity matrix, the problem (7.16) is simplified to the resource allocation problem (Yi et al. 2016). The allocation decisions are made to minimize the sum of all the agents' local objective functions while satisfying both the global network resource constraint and the local allocation feasibility constraints.

7.3.2 *Distributed Continuous-Time Algorithm Based on Proximal Operator and Gradient*

As mentioned above, the resource allocation problem can be viewed as distributed optimization with equality constraints. Compared with the existing optimal resource allocation (Deng et al. 2018; Yi et al. 2016; Zhu et al. 2021), the proposed proximal-gradient algorithm is not only smooth but can solve the nonsmooth optimization for second-order multi-agent systems. This section is divided into two parts: Firstly, smooth proximal-gradient algorithm is proposed for nonsmooth resource allocation problem of single-integrator multi-agent systems. Secondly, the proposed algorithm is extended to solve the optimal resource allocation problem for double-integrator multi-agent systems.

7.3.2.1 Optimal Resource Allocation for Single-Integrator Multi-agent Systems

For the single-integrator multi-agent systems with equality constraints as shown in (7.16), the following algorithm is proposed:

$$\begin{cases} \dot{x}_i = \text{prox}_{g_i}(x_i - \nabla f_i + w_i^T \eta_i) - x_i, \\ \dot{\eta}_i = -\sum_{j=1}^{n} a_{ij}(\eta_i - \eta_j) - \sum_{j=1}^{n} a_{ij}(z_i - z_j) + d_i - w_i x_i - w_i \dot{x}_i, \\ \dot{z}_i = \sum_{j=1}^{n} a_{ij}(\eta_i - \eta_j), \end{cases} \tag{7.17}$$

where $\eta_i \in \mathbb{R}^p$ and $z_i \in \mathbb{R}^p$. Denote $x \triangleq [x_1^T, x_2^T, \cdots, x_n^T]^T \in \mathbb{R}^{nq}$, $d \triangleq [d_1^T, d_2^T, \cdots, d_n^T]^T \in \mathbb{R}^{np}$, $z \triangleq [z_1^T, z_2^T, \cdots, z_n^T]^T \in \mathbb{R}^{np}$, $\eta \triangleq [\eta_1^T, \eta_2^T, \cdots, \eta_n^T]^T \in \mathbb{R}^{np}$, $L_p \triangleq L \otimes I_p \in \mathbb{R}^{np \times np}$, and $w \triangleq \text{diag}\{w_1, w_2, \cdots, w_n\} \in \mathbb{R}^{np \times nq}$.

Remark 7.6 A modified Lagrangian function $\mathcal{L}(x, z, \eta) = f(x) + g(x) + \eta^T(d - wx - L_p z) - 1/2\eta^T L_p \eta$ is selected corresponding to (7.17), where x and z are the primal variables and η represent the dual variables.

Remark 7.7 The main idea is to make a copy of the proximal term by introducing the derivative feedback so as to eliminate the "trouble" caused by the proximal term. The derivative term $\dot{x}_i$ on the right of the second equation in (7.17) is used to cancel the "trouble" term $\text{prox}_{g_i}(x_i - \nabla f_i + w_i^T \eta_i)$ in the analysis shown in (7.19)–(7.22) later.

Theorem 7.2 *Suppose Assumption 7.1 holds and consider algorithm* (7.17). *If positive parameters δ satisfy the inequalities*

$$m - \frac{\theta^2}{4\delta} > 0 \quad and \quad 0 < \delta \leq 1,$$

the distributed algorithm (7.17) *solves the optimization problem (7.16) for single-integrator multi-agent systems.*

Proof The compact form of algorithm (7.17) is

$$\begin{cases} \dot{x} = \text{prox}_g(x - \nabla f(x) + w^T \eta) - x, \\ \dot{\eta} = -L_p \eta - L_p z + d - wx - w\dot{x}, \\ \dot{z} = L_p \eta. \end{cases} \tag{7.18}$$

Assume the equilibrium point is $(x^*, \eta^*, z^*) \in \mathbb{R}^{nq\times np\times np}$. Similar to the Proof of Lemma 7.2, it is easy to prove x^* is the solution to problem (7.16).

Substituting the equilibrium point into (7.18), it yields

$$\begin{cases} x^* = \text{prox}_g(x^* - \nabla f(x^*) + w^T\eta^*), \\ d = wx^* + L_p z^*, \\ L_p\eta^* = \mathbf{0}_{np}. \end{cases} \tag{7.19}$$

Thus, it is derived that

$$\begin{aligned} &(x^* - \nabla f(x^*) + w^T\eta^*) - x^* \in \partial_g(x^*), \\ &(x - \nabla f(x) + w^T\eta) - x - \dot{x} \in \partial_g(x + \dot{x}). \end{aligned} \tag{7.20}$$

Recalling the property of proximal operator in (7.1), it is derived from (7.20)

$$(-(\nabla f(x) - \nabla f(x^*)) + w^T(\eta - \eta^*) - \dot{x})^T(x + \dot{x} - x^*) \geq 0. \tag{7.21}$$

Therefore, it yields

$$\begin{aligned} &-(\nabla f(x) - \nabla f(x^*))^T(x - x^*) + (\eta - \eta^*)^T w(x - x^*) - \dot{x}^T(x - x^*) - \\ &\qquad (\nabla f(x) - \nabla f(x^*))^T\dot{x} + (\eta - \eta^*)^T w\dot{x} - \dot{x}^T\dot{x} \geq 0. \end{aligned} \tag{7.22}$$

Choose the Lyapunov function candidate as

$$V_A = \frac{1}{2}\|x - x^*\|^2 + \frac{1}{2}\|\eta - \eta^*\|^2 + \frac{1}{2}\|z - z^*\|^2. \tag{7.23}$$

Denote $\hat{x} = x - x^*$, $\hat{\eta} = \eta - \eta^*$ and $\hat{z} = z - z^*$. Following (7.22), the derivative of V_A satisfies

$$\begin{aligned} \dot{V}_A =&(x - x^*)^T\dot{x} + (\eta - \eta^*)^T\dot{\eta} + (z - z^*)^T\dot{z} \\ \leq& - h^T(x - x^*) + (\eta - \eta^*)^T w(x - x^*) - h^T\dot{x} + (\eta - \eta^*)^T w\dot{x} \\ &- \dot{x}^T\dot{x} + (\eta - \eta^*)^T\dot{\eta} + (z - z^*)^T\dot{z} \\ \leq& - h^T(x - x^*) - h^T\dot{x} - \dot{x}^T\dot{x} - \hat{\eta}^T L_p\hat{\eta}, \end{aligned} \tag{7.24}$$

where $h = \nabla f(x) - \nabla f(x^*)$.

With the fact that $h^T(x - x^*) \geq m(x - x^*)^T(x - x^*)$ and $-h^T\dot{x} \leq \frac{1}{4\delta}h^T h + \delta\dot{x}^T\dot{x}$ $(\delta > 0)$, it is derived from (7.24) that

$$\begin{aligned}\dot{V}_A &\leq -m\hat{x}^T\hat{x} + \frac{1}{4\delta}h^T h + \delta\dot{x}^T\dot{x} - \dot{x}^T\dot{x} - \hat{\eta}^T L_p\hat{\eta} \\ &\leq -(m - \frac{\theta^2}{4\delta})\hat{x}^T\hat{x} - (1-\delta)\dot{x}^T\dot{x} - \hat{\eta}^T L_p\hat{\eta} \\ &\leq 0,\end{aligned} \tag{7.25}$$

where $m > \frac{\theta^2}{4\delta}$, $0 < \delta \leq 1$.

For all $(x, \eta, z) \in \mathbb{R}^{nq} \times \mathbb{R}^{np} \times \mathbb{R}^{np}$, V_A is positive definite. $V_A = 0$ if and only if $(x, \eta, z) = (x^*, \eta^*, z^*)$, and $V_A \to \infty$ as $(x, \eta, z) \to \infty$. Hence, the solution (x, η, z) is bounded for all $t > 0$. Let $\mathcal{Z} = \{(x, \eta, z) \in \mathbb{R}^{nq} \times \mathbb{R}^{np} \times \mathbb{R}^{np} : \dot{V}_A = 0\} \subset \{(x, \eta, z) \in \mathbb{R}^{nq} \times \mathbb{R}^{np} \times \mathbb{R}^{np} : (m - \frac{\theta^2}{4\delta})\hat{x}^T\hat{x} = 0,\ (1-\delta)\dot{x}^T\dot{x} = 0,\ \hat{\eta}^T L_p\hat{\eta} = 0\}$. Let $\mathcal{M}$ be the largest invariant subset of $\overline{\mathcal{Z}}$. It follows from Lemma 7.1 that $(x(t), \eta(t), z(t)) \to \mathcal{M}$ as $t \to \infty$. Note that $(m - \frac{\theta^2}{4\delta})\hat{x}^T\hat{x} > 0$ if $x \neq x^*$ because $m > \frac{\theta^2}{4\delta}$. Hence, $x(t) \to x^*$ as $t \to \infty$, implying that x converges to the optimal solution to problem (7.16). □

7.3.2.2 Optimal Resource Allocation for Double-Integrator Multi-agent Systems

Motivated by algorithm (7.17) for single-integrator multi-agent systems, a proximal-gradient algorithm is proposed for the distributed optimization of double-integrator multi-agent systems.

The proposed algorithm is described as:

$$\begin{cases} \dot{x}_i = y_i, \\ \dot{y}_i = \text{prox}_{g_i}(x_i - ky_i - \nabla f_i + w_i^T\eta_i) - x_i - \gamma y_i, \\ \dot{\eta}_i = -\sum_{j=1}^{n} a_{ij}(\eta_i - \eta_j) - \sum_{j=1}^{n} a_{ij}(z_i - z_j) + d_i - w_i x_i - w_i\dot{y}_i - \gamma w_i y_i, \\ \dot{z}_i = \sum_{j=1}^{n} a_{ij}(\eta_i - \eta_j), \end{cases} \tag{7.26}$$

where $y_i \in \mathbb{R}^q$.

Remark 7.8 Compared with algorithm (7.17), $\dot{y}_i$ is the derivative-feedback term in algorithm (7.26) and plays an important role in eliminating the "trouble" caused by the proximal term $\text{prox}_{g_i}(x_i - ky_i - \nabla f_i + w_i^T\eta_i)$ in the analysis shown in (7.30) and (7.31) later.

Theorem 7.3 *Suppose Assumption 7.1 holds and consider algorithm* (7.26). *If positive parameters k, γ, ϵ_1, ϵ_2 satisfy the inequalities*

$$\begin{cases} k \geq 1, \\ m - \dfrac{\gamma\theta^2}{4\epsilon_1} - \dfrac{\theta^2}{4\epsilon_2} > 0, \\ \gamma k + \gamma^2 - 1 - \gamma\epsilon_1 \geq 0, \\ 1 - \epsilon_2 \geq 0, \end{cases} \tag{7.27}$$

the distributed algorithm (7.26) *solves the optimization problem (7.16) for second-order multi-agent systems.*

Proof The compact form of algorithm (7.26) is

$$\begin{cases} \dot{x} = y, \\ \dot{y} = \text{prox}_g(x - ky - \nabla f(x) + w^T\eta) - x - \gamma y, \\ \dot{\eta} = -L_p\eta - L_p z + d - wx - w\dot{y} - \gamma wy, \\ \dot{z} = L_p\eta. \end{cases} \tag{7.28}$$

Assume the equilibrium point is (x^*, y^*, η^*, z^*). Similar to the Proof of Lemma 7.2, it is easy to prove x^* is the solution to problem (7.16).

Substituting the equilibrium point into (7.28), it yields

$$\begin{cases} \text{prox}_g(x^* - ky^* - \nabla f(x^*) + w^T\eta^*) = x^*, \\ d = wx^* + L_p\eta^* + L_p z^*, \\ L_p\eta^* = \mathbf{0}_{np}, \\ y^* = \mathbf{0}_{np}. \end{cases} \tag{7.29}$$

Furthermore, it follows that

$$\begin{aligned} &(x - ky - \nabla f(x) + w^T\eta) - x - \dot{y} - \gamma y \in \partial_g(x + \dot{y} + \gamma y), \\ &(x^* - ky^* - \nabla f(x^*) + w^T\eta^*) - x^* \in \partial_g(x^*). \end{aligned} \tag{7.30}$$

Finally, it is derived from (7.30) that

$$(-(k+\gamma)(y - y^*) - (\nabla f(x) - \nabla f(x^*)) + w^T(\eta - \eta^*) - \dot{y})^T(x - x^* + \dot{y} + \gamma y) \geq 0,$$

implying that

$$\begin{aligned} &-(k+\gamma)(y - y^*)^T(x - x^*) - (\nabla f(x) - \nabla f(x^*))^T(x - x^*) \\ &+ (\eta - \eta^*)^T w(x - x^*) - \dot{y}^T(x - x^*) - \gamma(k+\gamma)(y - y^*)^T(y - y^*) \\ &- \gamma(\nabla f(x) - \nabla f(x^*))^T(y - y^*) + \gamma(\eta - \eta^*)^T w(y - y^*) - \gamma\dot{y}^T(y - y^*) \\ &- (k+\gamma)(y - y^*)^T\dot{y} - (\nabla f(x) - \nabla f(x^*))^T\dot{y} + (\eta - \eta^*)^T w\dot{y} - \dot{y}^T\dot{y} \geq 0, \end{aligned} \tag{7.31}$$

To simplify (7.31), denote $\hat{x} = x - x^*$, $\hat{y} = y - y^* = y$, $\hat{\eta} = \eta - \eta^*$, $h = \nabla f(x) - \nabla f(x^*)$ and it yields

$$
\begin{aligned}
&-\gamma\hat{\eta}w\hat{y} + (k+2\gamma)\dot{\hat{y}}^T\hat{y} + (k+\gamma)\hat{y}^T\hat{x} - \hat{\eta}^T w\hat{x} + \dot{y}^T\hat{x} - \hat{\eta}^T w\dot{y} \\
&\leq -\gamma(k+\gamma)\hat{y}^T\hat{y} - \gamma h^T\hat{y} - h^T\hat{x} - h^T\dot{y} - \dot{y}^T\dot{y}
\end{aligned}
\tag{7.32}
$$

Choose the Lyapunov function candidate

$$
\begin{aligned}
V_B =& \frac{k+\gamma}{2}\|x - x^*\|^2 + \frac{k+2\gamma}{2}\|y - y^*\|^2 + (x - x^*)^T(y - y^*) \\
&+ \frac{1}{2}\|\eta - \eta^*\|^2 + \frac{1}{2}\|z - z^*\|^2.
\end{aligned}
\tag{7.33}
$$

Obviously, V_B is positive definite when $k \geq 1$. With the fact $d = wx^* + L_p\eta^* + L_p z^*$ in (7.29), the derivative of V_B is derived as follows

$$
\begin{aligned}
\dot{V}_B &= (k+\gamma)\hat{x}^T\dot{\hat{x}} + (k+2\gamma)\hat{y}^T\dot{\hat{y}} + \hat{x}^T\dot{y} + \hat{y}^T\hat{y} + \hat{z}^T\dot{\hat{z}} + \hat{\eta}^T\dot{\hat{\eta}} \\
&= (k+\gamma)\hat{x}^T\hat{y} + (k+2\gamma)\hat{y}^T\dot{\hat{y}} + \hat{x}^T\dot{y} + \hat{y}^T\hat{y} + \hat{z}^T L_p\eta \\
&+ \hat{\eta}^T(d - wx - L_p z - L_p\eta - w\dot{y} - \gamma wy) \\
&= (k+\gamma)\hat{x}^T\hat{y} + (k+2\gamma)\hat{y}^T\dot{\hat{y}} + \hat{x}^T\dot{y} + \hat{y}^T\hat{y} + \hat{z}^T L_p\hat{\eta} \\
&+ \hat{\eta}^T(-w\hat{x} - L_p\hat{z} - L_p\hat{\eta} - w\dot{y} - \gamma wy).
\end{aligned}
\tag{7.34}
$$

Combining (7.32) and (7.34), it has

$$
\dot{V}_B \leq -\hat{\eta}^T L_p\hat{\eta} - \gamma(k+\gamma)\hat{y}^T\hat{y} + \hat{y}^T\hat{y} - h^T\hat{x} - h^T\dot{y} - \gamma h^T\hat{y} - \dot{y}^T\dot{y}. \tag{7.35}
$$

With the facts that $h^T\hat{x} \geq m\hat{x}^T\hat{x}$, $-\gamma h^T\hat{y} \leq \frac{\gamma}{4\epsilon_1}h^T h + \gamma\epsilon_1\hat{y}^T\hat{y}$ $(\epsilon_1 > 0)$, $-h^T\dot{y} \leq \frac{1}{4\epsilon_2}h^T h + \epsilon_2\dot{y}^T\dot{y}$ $(\epsilon_2 > 0)$ and $h^T h \leq \theta^2\hat{x}^T\hat{x}$, it is obtained that

$$
\begin{aligned}
\dot{V}_B &\leq -\hat{\eta}^T L_p\hat{\eta} - (\gamma k + \gamma^2 - 1)\hat{y}^T\hat{y} - m\hat{x}^T\hat{x} - h^T\dot{y} - \gamma h^T\hat{y} - \dot{y}^T\dot{y} \\
&\leq -\hat{\eta}^T L_p\hat{\eta} - \mu_1\hat{x}^T\hat{x} - \mu_2\hat{y}^T\hat{y} - \mu_3\dot{y}^T\dot{y}, \\
&\leq 0,
\end{aligned}
\tag{7.36}
$$

where $\mu_1 = m - \frac{\gamma\theta^2}{4\epsilon_1} - \frac{\theta^2}{4\epsilon_2}$, $\mu_2 = \gamma k + \gamma^2 - 1 - \gamma\epsilon_1$ and $\mu_3 = 1 - \epsilon_2$.

For all $(x, y, \eta, z) \in \mathbb{R}^{nq} \times \mathbb{R}^{nq} \times \mathbb{R}^{np} \times \mathbb{R}^{np}$, V_B is positive definite. $V_B = 0$ if and only if $(x, y, \eta, z) = (x^*, y^*, \eta^*, z^*)$, and $V_B \to \infty$ as $(x, y, \eta, z) \to \infty$. Let $\mathcal{Z} = \{(x, y, \eta, z) \in \mathbb{R}^{nq} \times \mathbb{R}^{nq} \times \mathbb{R}^{np} \times \mathbb{R}^{np} : \dot{V}_B = 0\} \subset \{(x, y, \eta, z) \in \mathbb{R}^{nq} \times \mathbb{R}^{nq} \times \mathbb{R}^{np} \times \mathbb{R}^{np} : \mu_1\hat{x}^T\hat{x} = 0,\ \mu_2\hat{y}^T\hat{y} = 0,\ \mu_3\dot{y}^T\dot{y} = 0,\ \hat{\eta}^T L_p\hat{\eta} = 0\}$. Let $\mathcal{M}$ be the largest invariant subset of $\overline{\mathcal{Z}}$. It follows from Lemma 7.1 that $(x(t), y(t), \eta(t), z(t))$

$\to \mathcal{M}$ as $t \to \infty$. Note that $\mu_1 \hat{x}^T \hat{x} > 0$ if $x \neq x^*$ because $\mu_1 > 0$. Hence, $x(t) \to x^*$ as $t \to \infty$, implying that the distributed optimization problem (7.16) is solved. □

Remark 7.9 In this section, different from the algorithm (7.13) for consensus problem, an auxiliary variable z_i is introduced in algorithms (7.17) and (7.26), which plays an important role to make the dual variable η_i reach consensus.

7.4 Numerical Simulation

In this section, we give two numerical examples to illustrate the effectiveness of the proposed algorithms for optimal consensus and resource allocation problems of second-order multi-agent systems, respectively. Consider a connected and undirected multi-agent system grouped by six agents, which is shown in Fig. 7.1.

7.4.1 An Example for Optimal Consensus

Consider the following nonsmooth optimal consensus problem

$$\begin{aligned} \min \quad & \sum_{i=1}^{6} (f_i(x_i) + g_i(x_i)), \\ \text{subject to} \quad & x_i = x_j, \quad \forall i, j \in \{1, ..., 6\}, \end{aligned}$$

where $x_i \in \mathbb{R}$ and the local cost functions are given by

$$\begin{aligned} g_i(x_i) &= \|x_i\|_1, \\ f_1(x_1) &= \|x_1 - 4\|^2, \\ f_2(x_2) &= \|x_2 - 2\|^2 + x_2, \\ f_3(x_3) &= \|x_3\|^2 + 3, \end{aligned}$$

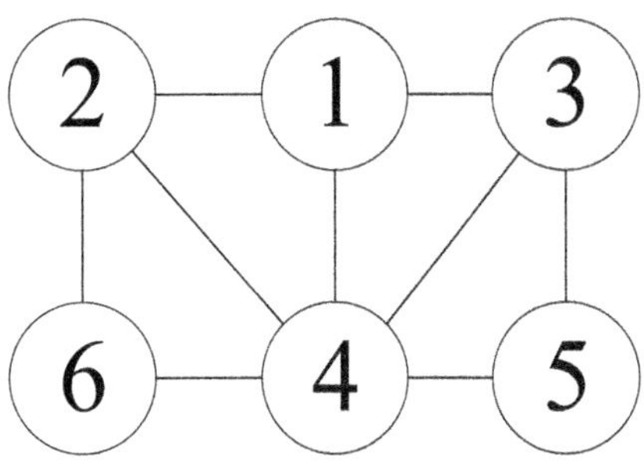

Fig. 7.1 The communication topology

Fig. 7.2 The trajectories of the states x_i and y_i $(i = 1, 2, \cdots, 6)$

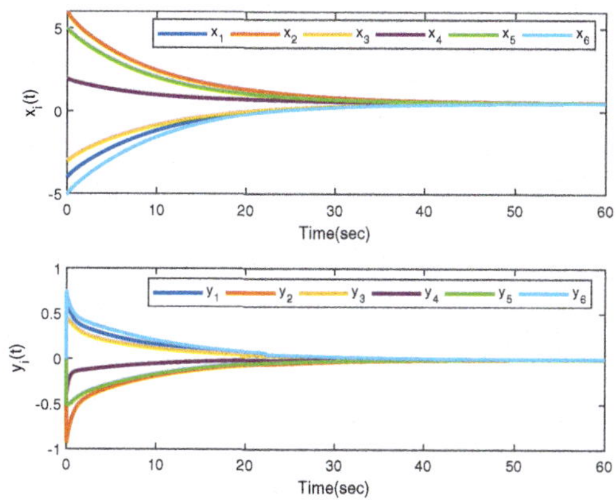

$$f_4(x_4) = \|x_4 - 1\|^2,$$
$$f_5(x_5) = \|x_5 + 1\|^2 - x_5,$$
$$f_6(x_6) = \|x_6 - 2\|^2.$$

$f_i(x_i)$ and $g_i(x_i)$ represent the quadratic objective and the l_1 penalty with the anchor 0 for each agent i, respectively (Wei et al. 2020).

Obviously, $g_i(x)$ $(i = 1, 2, \cdots, 6)$ is non-differentiable. $f_i(x)$ $(i = 1, 2, \cdots, 6)$ is m_i-strongly convex and its gradient satisfies θ_i-Lipschitz condition. We select $\theta = 2$, $m = 2$. The maximum eigenvalue of the Laplacian matrix corresponding to Fig. 7.1 is 6. Solve the above optimal consensus problem with algorithm (7.13) for second-order multi-agent systems and select $\beta = 80$, $\gamma = 10$, $c = 3$ so that the inequalities in Theorem 1 are satisfied. The initial states for six agents are set as $x(0) = [-4; 6; -3; 2; 5; -5]$, initial y and η are zero vectors, respectively. Figure 7.2 shows the trajectories of states x_i and y_i for the six agents, respectively. From Fig. 7.2, it is easy to see that the states of the six agents finally reach consensus. Figure 7.3 shows the dual variable η_i for each agent. Figure 7.4 describes the trajectory of the objective function, denoted by $f(x) + g(x)$ for short. Meanwhile, the minimum value 25 of the objective function satisfying consensus constraint is obtained by theoretic calculation, which is denoted by the dashed line in Fig. 7.4.

Select the Lagrangian function $\mathcal{L}(x, \eta) = f(x) + g(x) + \eta^T L_q x$, where x is the primal variable and η is the dual variable. Here we give the trajectories of $\partial_x \mathcal{L}(x, \eta) = \nabla f(x) + \partial g(x) + L_q \eta$ and $\nabla_\eta \mathcal{L}(x, \eta) = L_q x$ as shown in Fig. 7.5. $\mathcal{N}(\partial_x \mathcal{L}(x, \eta))$ denotes the element of $\partial_x \mathcal{L}(x, \eta)$ with minimal norm. From Fig. 7.5, it follows that $\mathcal{N}(\partial_x \mathcal{L}(x(t), \eta(t))) \to 0$ and $\nabla_\eta \mathcal{L}(x(t), \eta(t)) \to 0$ as $t \to \infty$, implying the optimal solution is achieved.

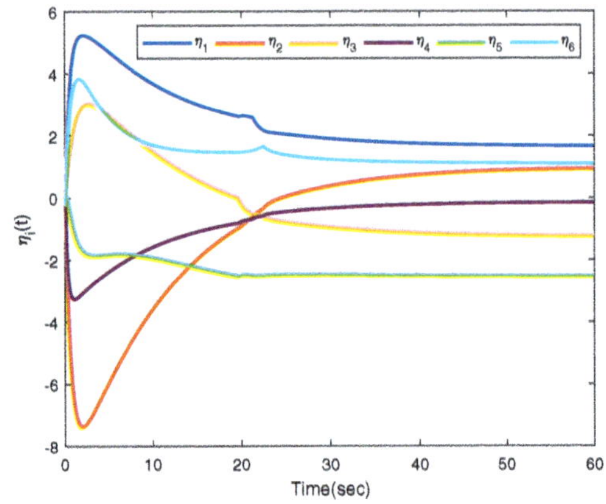

Fig. 7.3 The trajectories of dual variables η_i $(i = 1, 2, \cdots, 6)$

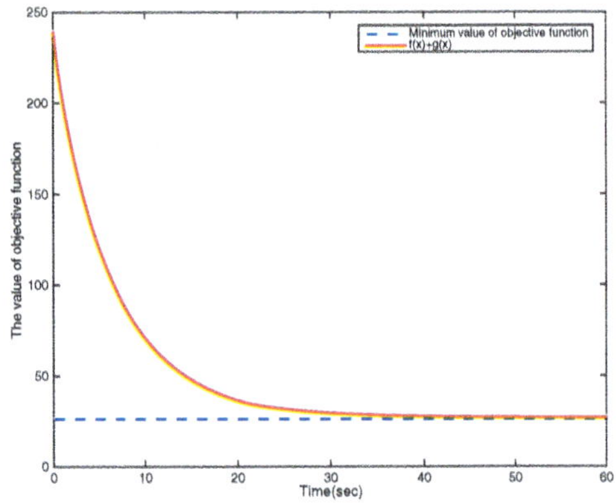

Fig. 7.4 The trajectory of the objective function $f(x) + g(x)$

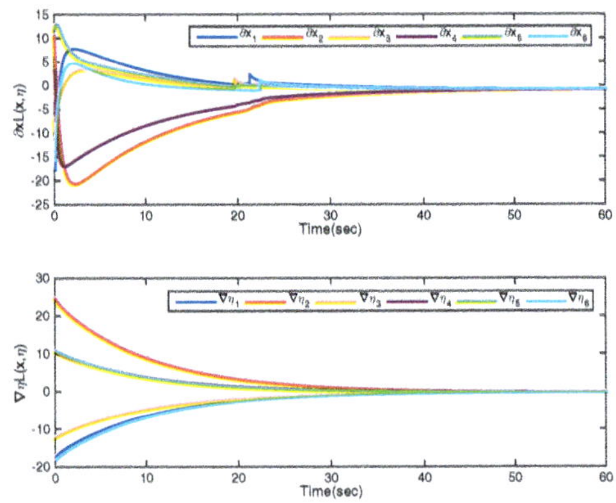

Fig. 7.5 $\partial_x \mathcal{L}(x, \eta)$ and $\nabla_\eta \mathcal{L}(x, \eta)$ of Lagrangian function $\mathcal{L}(x, \eta)$

7.4.2 An Example for Optimal Resource Allocation

Consider the following nonsmooth distributed optimization for resource allocation problem (Zeng et al. 2018)

$$\begin{aligned} \min \quad & \sum_{i=1}^{6}(f_i(x_i)+g_i(x_i)), \\ \text{subject to} \quad & \sum_{i=1}^{6} x_i = \sum_{i=1}^{6} d_i, \end{aligned}$$

where $x_i \in \mathbb{R}^2, i \in \{1, \cdots, 6\}$ and $d_i = [0.3; 0.2]$. The local cost functions are given by

$$\begin{aligned} g_i(x_i) &= \begin{cases} 0, & x_i \in \Omega_i, \\ \infty, & x_i \notin \Omega_i, \end{cases} \\ f_i(x_i) &= \|x_i - p_i\|^2, \end{aligned}$$

where $\Omega_i = \{\delta \in \mathbb{R}^2 | \|\delta - x_i(0)\|^2 \le 64\}$, $p_i = [i-2, 0]^T$.

$f_i(x_i)$ and $g_i(x_i)$ represent the quadratic objective, the indicator function of the constraint set $x_i \in \Omega_i$, respectively. Based on the definition above, the proximal operator and gradient for each agent are $\text{prox}_{g_i}(v) = \arg\min_{\delta\in\Omega_i} \|\delta - v\|^2$ and $\nabla f_i(x_i) = 2(x_i - m_i) \in \mathbb{R}^2$, respectively. The maximum eigenvalue of the Laplacian matrix corresponding to Fig. 7.1 is 6.

The algorithm (7.26) is designed to solve the resource allocation problem for the second-order multi-agent system. Select parameters $k = 5$, $\gamma = 0.5$, $\epsilon_1 = \epsilon_2 = 1$. The initial values $x_i(0)$ $(i = 1, \cdots, 6)$ are given as $x_1(0) = [-4; 5.5]$, $x_2(0) = [6; 5]$, $x_3(0) = [-3; 2]$, $x_4(0) = [2; -3]$, $x_5(0) = [5; -3.5]$, $x_6(0) = [-5; 5]$, $y_i(0)$ and $\eta_i(0)$ are zero vectors. Figures 7.6, 7.7 and 7.8 show the trajectories of the generalized coordinates x and y, the auxiliary variable η of the six agents with algorithm (7.26), respectively. Figure 7.9 shows the errors of the constraint condition $\sum_{i=1}^{n} x_i - \sum_{i=1}^{n} d_i$, denoted by $[e_1\ e_2]^T$ for short. Figure 7.10 shows the trajectory of the summation of the local objective functions, where the dashed line denotes the real minimum value 8.64 of the objective function satisfying equality constraint. Through the simulations, we can see that the errors can converge to zero, which means the constraint condition $\sum_{i=1}^{n} x_i = \sum_{i=1}^{n} d_i$ is satisfied.

Select the Lagrangian function $\mathcal{L}(x, z, \eta) = f(x) + g(x) + \eta^T(d - wx - L_p z) - 1/2\eta^T L_p \eta$, where x, z are the primal variables and η is the dual variable. Here we give the trajectories of $\partial_x \mathcal{L}(x, z, \eta) = \nabla f(x) + \partial g(x) + w^T\eta$, $\nabla_z \mathcal{L}(x, z, \eta) =$

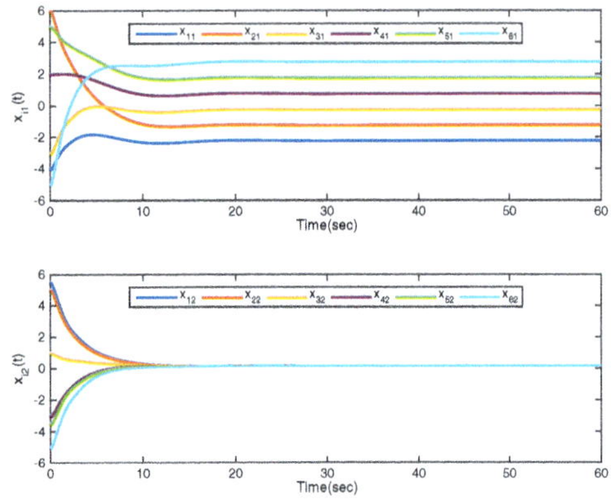

Fig. 7.6 The trajectories of the states x_i $(i = 1, 2, \cdots, 6)$

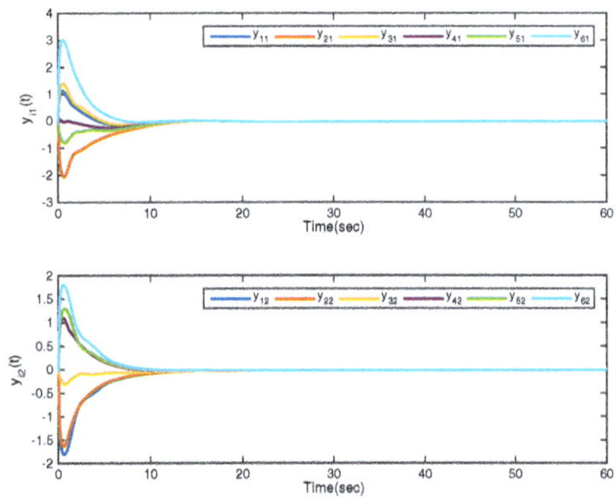

Fig. 7.7 The trajectories of the states y_i $(i = 1, 2, \cdots, 6)$

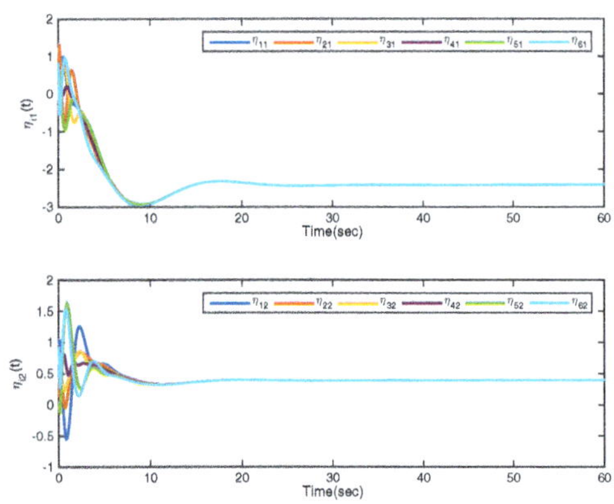

Fig. 7.8 The trajectories of dual variables η_i $(i = 1, 2, \cdots, 6)$

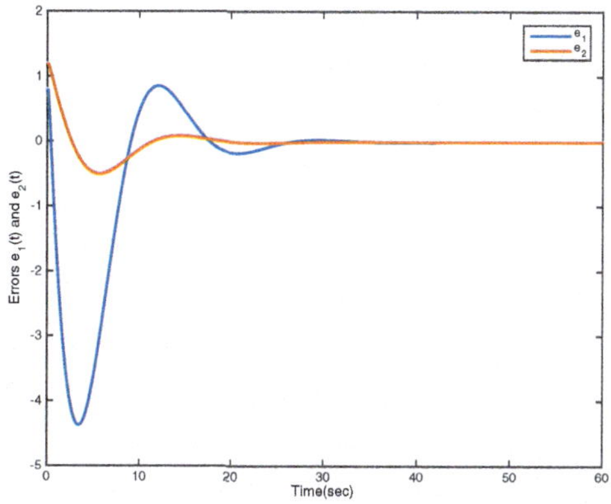

Fig. 7.9 The errors e_1 and e_2

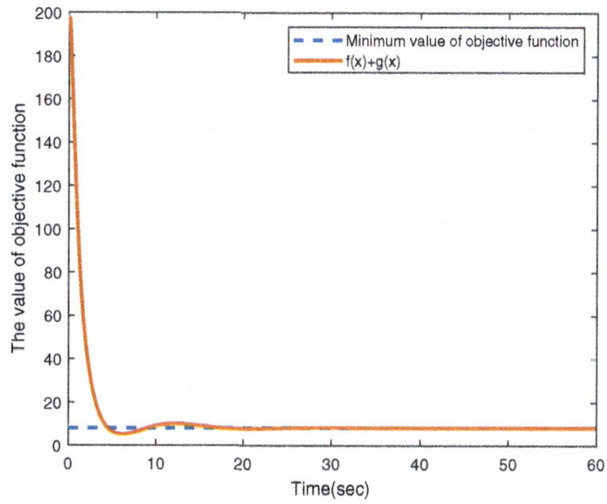

Fig. 7.10 The trajectory of the objective function $f(x) + g(x)$

$-L_p\eta$ and $\nabla_\eta \mathcal{L}(x, z, \eta) = d - wx - L_p z - L_p \eta$ as shown in Fig. 7.11. $\mathcal{N}(\partial_x \mathcal{L}(x, z, \eta))$ denotes the element of $\partial_x \mathcal{L}(x, z, \eta)$ with minimal norm. From Fig. 7.11, it follows that $\mathcal{N}(\partial_x \mathcal{L}(x(t), z(t), \eta(t))) \to 0$, $\nabla_z \mathcal{L}(x(t), z(t), \eta(t)) \to 0$ and $\nabla_\eta \mathcal{L}(x(t), z(t), \eta(t)) \to 0$ as $t \to \infty$, implying the optimal solution is achieved.

7.5 Conclusion

In this chapter, the distributed nonsmooth convex optimization is investigated for second-order multi-agent systems. Based on the proximal operator and derivative feedback techniques, the proximal-gradient algorithms are proposed to solve the distributed optimal consensus and resource allocation problems, respectively. The

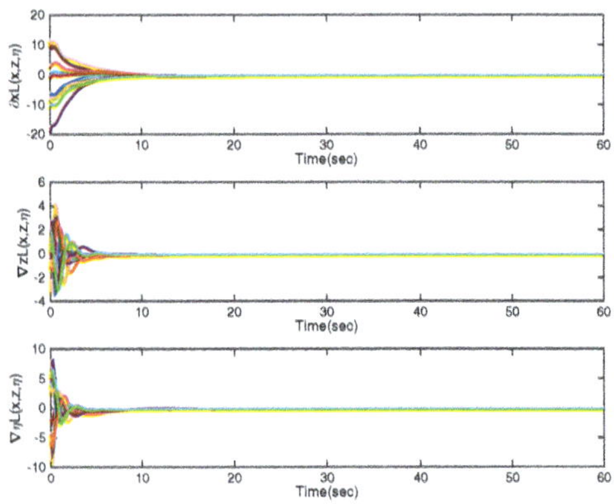

Fig. 7.11 $\partial_x \mathcal{L}(x, z, \eta)$, $\nabla_z \mathcal{L}(x, z, \eta)$ and $\nabla_\eta \mathcal{L}(x, z, \eta)$ of Lagrangian function $\mathcal{L}(x, z, \eta)$

proposed algorithms can converge to the optimal solution as well as satisfy the equality constraints. Future research includes the proof of convergence rate and extending results to high-order or nonlinear multi-agent systems with directed topologies.

References

Boyd, S., Parikh, N., Chu, E., Peleato, B., and Eckstein, J. (2010). "Distributed optimization and statistical learning via the alternating direction method of multipliers," *Foundations and Trends in Machine Learning* **3**, 1, pp. 1–122.

Burger, M., Notarstefano, G., Bullo, F., and Allgower, F. (2012). "A distributed simplex algorithm for degenerate linear programs and multi-agent assignments," *Automatica* **48**, 9, pp. 2298–2304.

Carlone, L., Srivastava, V., Bullo, F., and Calafiore, G. C. (2014). "Distributed random convex programming via constraints consensus," *SIAM Journal on Control and Optimization* **52**, 1, pp. 629–662.

Deng, Z. H., Liang, S., and Yu, W. Y. (2018). "Distributed optimal resource allocation of second-order multiagent systems," *International Journal of Robust and Nonlinear Control* **28**, 14, pp. 4246–4260.

Gharesifard, B. and Cortes, J. (2014). "Distributed continuous-time convex optimization on weight-balanced digraphs," *IEEE Transactions on Automatic Control* **59**, 3, pp. 781–786.

Hui, Q., Haddad, W. M., and Bhat, S. P. (2009). "Semistability, finite-time stability, differential inclusions, and discontinuous dynamical systems having a continuum of equilibria," *IEEE Transactions on Automatic Control* **54**, 10, pp. 2465–2470.

Li, X. X., Xie, L. H., and Hong, Y. G. (2019). "Distributed continuous-time algorithm for a general nonsmooth monotropic optimization problem," *International Journal of Robust and Nonlinear Control* **29**, 10, pp. 3252–3266.

Liang, S., Zeng, X. L., and Hong, Y. G. (2018). "Distributed nonsmooth optimization with coupled inequality constraints via modified lagrangian function," *IEEE Transactions on Automatic Control* **63**, 6, pp. 1753–1759.

Lin, P., Ren, W., and Farrell, J. A. (2017). "Distributed continuous-time optimization: Nonuniform gradient gains, finite-time convergence, and convex constraint set," *IEEE Transactions on Automatic Control* **62**, 5, pp. 2239–2253.

Liu, Q. S. and Wang, J. (2013). "A one-layer projection neural network for nonsmooth optimization subject to linear equalities and bound constraints," *IEEE Transactions on Neural Networks and Learning Systems* **24**, 5, pp. 812–824.

Nedic, A., Ozdaglar, A., and Parrilo, P. A. (2010). "Constrained consensus and optimization in multi-agent networks," *IEEE Transactions on Automatic Control* **55**, 4, pp. 922–938.

Ozdaglar, D. P. B. A. (2003). *Convex Analysis and Optimization* (Athena Scientific).

Parikh, N. and Stephen, B. (2014). "Proximal algorithms," *Foundations and Trends extregistered in Optimization* **3**, 1, pp. 127–239.

Ram, S. S., Nedic, A., and Veeravalli, V. V. (2010). "Distributed stochastic subgradient projection algorithms for convex optimization," *Journal of Optimization Theory and Applications* **147**, 3, pp. 516–545.

Ruszczyski, A. (2006). *Nonlinear Optimization* (Princeton University Press).

Shi, G. D., Johansson, K. H., and Hong, Y. G. (2013). "Reaching an optimal consensus: Dynamical systems that compute intersections of convex sets," *IEEE Transactions on Automatic Control* **58**, 3, pp. 610–622.

Shor, N. Z. (2012). *Minimization methods for non-differentiable functions* (Springer Science & Business Media).

Tibshirani, R., Saunders, M., Rosset, S., Zhu, J., and Knight, K. (2005). "Sparsity and smoothness via the fused lasso," *Journal of the Royal Statistical Society Series B-Statistical Methodology* **67**, pp. 91–108.

Wei, Y., Fang, H., Zeng, X. L., Chen, J., and Pardalos, P. (2020). "A smooth double proximal primal-dual algorithm for a class of distributed nonsmooth optimization problems," *IEEE Transactions on Automatic Control* **65**, 4, pp. 1800–1806.

Yi, P., Hong, Y. G., and Liu, F. (2015). "Distributed gradient algorithm for constrained optimization with application to load sharing in power systems," *Systems & Control Letters* **83**, pp. 45–52.

Yi, P., Hong, Y. G., and Liu, F. (2016). "Initialization-free distributed algorithms for optimal resource allocation with feasibility constraints and application to economic dispatch of power systems," *Automatica* **74**, pp. 259–269.

Zeng, X. L., Yi, P., Hong, Y. G., and Xie, L. H. (2018). "Distributed continuous-time algorithms for nonsmooth extended monotropic optimization problems," *SIAM Journal on Control and Optimization* **56**, 6, pp. 3973–3993.

Zhang, Y. and Hong, Y. (2015). "Distributed optimization design for high-order multi-agent systems," in *2015 34th Chinese Control Conference (CCC)*, 2015 34th Chinese Control Conference (CCC), pp. 7251–7256.

Zhu, Y. N., Ren, W., Yu, W. W., and Wen, G. H. (2021). "Distributed resource allocation over directed graphs via continuous-time algorithms," *IEEE Transactions on Systems Man Cybernetics-Systems* **51**, 2, pp. 1097–1106.

Chapter 8
Distributed Optimal Consensus for Euler-Lagrange Systems Based on Event-Triggered Control

The Euler-Lagrange (EL) system is prevailing in the modeling and design for a large class of nonlinear systems, such as robotic manipulators, spacecraft and marine vessels. As a result, the optimal consensus of EL multi-agent systems has been widely studied. For example, two different distributed algorithms were proposed for heterogeneous continuous EL multi-agent systems to accomplish the optimization task in Zhang et al. (2017). In Deng and Yiguang (2016), distributed algorithms for EL systems with semi-global stability were developed. In Qiu and Yiguang (2016), distributed constrained consensus algorithms were developed for heterogeneous EL multi-agent systems with kinematic constraints.

In practical systems, limited energy and communication bandwidth are inevitable. It is meaningful to propose event-triggered strategies to avoid continuous communication among agents (Dimarogonas et al. 2012; Liu et al. 2018; Tran et al. 2019; Liu et al. 2019; Wu et al. 2018; Zhang et al. 2021; Liu et al. 2019a). Up to now, the majority of event-triggered optimal problems have been involved in single- and double-integrator dynamics (Yi et al. 2018; Deng et al. 2017; Li et al. 2018; Zhao et al. 2018; Shi et al. 2018). In Yi et al. (2018), a distributed algorithm based on event-triggered scheme for double-integrator multi-agent systems was developed to achieve distributed optimization. In Deng et al. (2017), a distributed optimization algorithm was proposed for continuous-time multi-agent systems with discrete-time communication and gradient measurement. In Li et al. (2018), a quantized event-triggered algorithm was considered to solve a large class of convex optimization problems for a set of single-integrator agents. In Zhao et al. (2018), a distributed event-triggered algorithm was developed to solve the convex quadratic optimization problem for a group of single-integrator multi-agent systems under undirected and connected topologies. In Shi et al. (2018), a distributed sub-gradient algorithm based on event-triggered scheme was developed for the resource allocation problem (RAP) with constraints under fixed directed topology.

There are a few results about optimal coordination of EL systems with event-triggered schemes such as Liu et al. (2019b) and Wang et al. (2017). In Liu et al.

Q. Wang et al., *Distributed Cooperative Control and Optimization for Multi-Agent Systems*, https://doi.org/10.1007/978-981-96-0950-5_8

(2019b), three different distributed event-triggered control algorithms were developed to achieve leader-follower consensus for the EL multi-agent systems, but continuous communication was still inevitable when the triggering condition was monitored. In Wang et al. (2017), an event-based continuous-time algorithm for the EL system was developed to achieve distributed optimal consensus when the model parameters of the EL system were available. However, there are some limitations that the algorithms monitor the triggering function continuously or rely on the model parameters of the EL system. On one hand, the effectiveness of the event-triggered scheme can be improved by monitoring the triggering function discretely. On the other hand, it is usually difficult to obtain the model parameters exactly due to the uncertainties, so it is practical to study the optimization problem for EL systems with unavailable model parameters.

In view of these reasons, the objective of this chapter is to develop distributed optimization algorithms based on event-triggered scheme to deal with the optimal consensus problem of EL multi-agent systems with lower communication and computation cost. The main contributions and innovations of this chapter are listed as follows:

- Firstly, a new distributed event-triggered algorithm is developed for the EL systems with available model parameters. Compared with the related work in Wang et al. (2017), the proposed algorithm is independent of neighbors' velocities, which can reduce communication and computation greatly.
- Secondly, when the model parameters are unavailable, a virtual system is introduced as a reference model and tracking controller is designed to follow the virtual system. Then the optimal consensus algorithm based on the event-triggered scheme is developed to achieve consensus as well as minimize the global cost function.
- Thirdly, compared with the algorithms in Zhang et al. (2017) and Liu et al. (2019b), continuous communication can be avoided during the process of triggering condition monitoring and update of controllers with the developed event-triggered scheme in the paper.
- Finally, in contrast to the semi-global convergence in Deng and Yiguang (2016) and the bounded stabilization in Kia et al. (2015), the developed algorithm in this paper can ensure the system to converge to the exact optimal solution exponentially.

8.1 Mathematical Preliminaries

8.1.1 Convex Definition

Convex set: A set $S \subseteq \mathbb{R}^m$ is called convex if for any $0 \leq \mu \leq 1$, $x + (1-\mu)y \in K$ is satisfied when $x, y \in S$.

Convex function: If for any $0 \leq \mu \leq 1$, $f(\mu x + (1-\mu)y) \leq \mu f(x) + (1-\mu)f(y)$ is satisfied when $x, y \in S$, the function $f(\cdot) : S \to \mathbb{R}$ is called convex, in which S is a convex set.

Strongly convex: Assume f is a differentiable function. If for any $x, y \in S$, $(\nabla f(x) - \nabla f(y))^T(x - y) \geq \omega\|x - y\|^2$ is satisfied, the function f is called convex. Furthermore, if for any $x, y \in S$, $(\nabla f(x) - \nabla f(y))^T(x - y) \geq \omega\|x - y\|^2$ is satisfied, f is called ω-strongly convex.

θ-Lipschitz: If for any $x, y \in S$, there exists constant $\theta > 0$ satisfying $\|f(x) - f(y)\| \leq \theta\|x - y\|$, the function $f : \mathbb{R}^m \to \mathbb{R}$ is called θ-Lipschitz over a set $S \in \mathbb{R}^m$.

8.1.2 Graph Theory

For a graph $\mathcal{G}$ with n agents, it is denoted by $\mathcal{G} = \{\mathcal{V}, \mathcal{E}\}$, where $\mathcal{V} = \{v_1, v_2, \ldots, v_n\}$ is the set of the nodes, and $\mathcal{E} = \{(v_i, v_j)|v_i, v_j \in \mathcal{V}\}$ is the set of edges. $\mathcal{G}$ is called undirected if $(v_i, v_j) \in \mathcal{E}$ as well as $(v_j, v_i) \in \mathcal{E}$. Meanwhile, it is said that node v_i is adjacent to node v_j if there exists an edge between v_i and v_j. The elements of adjacency matrix $A = [a_{ij}] \in \mathbb{R}^{n\times n}$ corresponding to $\mathcal{G}$ are given as $a_{ii} = 0, a_{ij} = 1$ if $(v_j, v_i) \in \mathcal{E}$, otherwise, $a_{ij} = 0$. $D = diag\{d_1, d_2, \ldots, d_n \in R^{n\times n}\}$ denotes the degree matrix of $\mathcal{G}$, where $d_i = \sum_{j=1}^{n} a_{ij}$ for $i = 1, \ldots, n$ and $d = \max_{i=1,2,\ldots,n} \{d_i\}$. $L = D - A$ denotes the Laplacian matrix of $\mathcal{G}$.

Obviously, for an undirected graph $\mathcal{G}$, it is easy to know that L is symmetric and positive semi-definite, i.e., $L^T = L \geq 0$. Meanwhile, the row sum of L is equal to zero and 0 is one of the eigenvalues. Therefore, the corresponding eigenvector associated with the eigenvalue 0 is the vector $\mathbf{1}$ and it has $L\mathbf{1} = \mathbf{0}$. For the connected graph $\mathcal{G}$, there exists exactly a zero eigenvalue corresponding to L, and all the eigenvalues can be sorted in an ascending order $0 = \lambda_1 < \lambda_2 \leq \cdots \leq \lambda_{\max}$.

8.1.3 Notations

For simplicity, $\mathbb{R}$ represents the set of real numbers; $\mathbb{R}^m$ represents the set of m-dimensional real column vectors; $\mathbb{R}^{n\times m}$ represents the set of $n \times m$ real matrices; $\mathbf{0}$ is adopted to denote zero matrices with appropriate dimension as well as zero vectors and zero constants; I_n stands for the identity matrix with appropriate dimension; $\|\cdot\|$ represents the Euclidean norm of a matrix. Furthermore, $(\cdot)^T$ represents the transpose of a matrix; $A \otimes B$ refers to the Kronecker product of matrices A and B.

$$\text{prox}_f(v) = \arg\min_x (f(x) + 1/2\|x - v\|^2).$$

8.2 Problem Formulation

Consider a group of n mechanical robots, each of which is described by an Euler-Lagrange equation as follows

$$M_i(q_i)\ddot{q}_i + C_i(q_i, \dot{q}_i)\dot{q}_i + G_i(q_i) = \tau_i, \quad i = 1, \ldots, n, \tag{8.1}$$

where $q_i, \dot{q}_i \in \mathbb{R}^m$ denote the generalized angle and angular velocity vectors, respectively; $M_i(q_i) \in \mathbb{R}^{m\times m}$ denotes the symmetric positive-definite inertia matrix; $C_i(q_i, \dot{q}_i)\dot{q}_i \in \mathbb{R}^m$ denotes the Coriolis and centrifugal torque, and $\tau_i \in \mathbb{R}^m$ denotes the control torque on robot i.

As well known that the Euler-Lagrange system has the following properties:
(1) $\dot{M}_i(q_i) - 2C_i(q_i, \dot{q}_i)$ is skew symmetric;
(2) For any $x, y \in \mathbb{R}^m$, $M_i(q_i)x + C_i(q_i, \dot{q}_i)y + G_i(q_i) = \Omega_i(q_i, \dot{q}_i, x, y)\varphi_i$, where $\Omega_i(q_i, \dot{q}_i, x, y) \in \mathbb{R}^{m\times p}$ is a known regression matrix and $\varphi_i \in \mathbb{R}^p$ is a constant vector and consists of the uncertain model parameters in systems (8.1).

As for the EL system (8.1), each agent i is presented with a local cost function $f_i : \mathbb{R}^m \to \mathbb{R}$, which is only known by itself. The objective of the system (8.1) is to achieve distributed optimal consensus, which is described as follows:

$$\min_{z\in\mathbb{R}^m} f(z), \quad f(z) = \sum_{i=1}^{n} f_i(z), \tag{8.2}$$

where $f(z)$ is defined as the global cost function of the whole system. Specifically, a strict definition is given as follows:

Definition 8.1 The distributed optimization is achieved for the system (8.1), if for any initial condition $q_i(0), \dot{q}_i(0) \in \mathbb{R}^m$, all the Euler-Lagrange agents can converge to the global optimal solution of problem (8.2), i.e.,

$$\lim_{t\to\infty} q_i(t) = z^*, \ \lim_{t\to\infty} \dot{q}_i(t) = 0_m, \tag{8.3}$$

where $z^* = \arg\min_{z\in\mathbb{R}^m} f(z)$.

Assumption 8.1 Graph $\mathcal{G}$ is undirected and connected.

Assumption 8.2 For each agent i, the function f_i is differentiable and ω_i-strongly convex satisfying $(a-b)^T(\nabla f_i(a) - \nabla f_i(b)) \ge \omega\|a-b\|^2, \forall a, b \in \mathbb{R}^m$.

Assumption 8.3 For each agent i, there exists a constant $\theta_i > 0$, such that $\|\nabla f_i(a) - \nabla f_i(b)\| \le \theta_i\|a-b\|, \forall a, b \in \mathbb{R}^m$.

Remark 8.1 The objective in this chapter is to design a control protocol τ_i to drive q_i to z^* with the reduction of communication cost so that the optimization problem $z^* = \arg\min_{z\in\mathbb{R}^m} f(z)$ is solved.

8.3 Event-Triggered Optimal Consensus with Available Model Parameters for EL Systems

In this subsection, the model parameters of the EL multi-agent systems are assumed to be available and the event-triggered optimal consensus protocol are proposed as follows

$$\begin{cases} \dot{q}_i = p_i, \\ \tau_i = C_i(q_i, \dot{q}_i)\dot{q}_i + G_i(q_i) \\ \qquad - kM_i(q_i)\dot{q}_i - \beta M_i(q_i) \sum\limits_{j \in N_i} a_{ij}(q_i(t_k^i) - q_j(t_{k'}^j)) \\ \qquad - M_i(q_i)\nabla f_i(q_i) - M_i(q_i)w_i, \\ \dot{w}_i = \beta \sum\limits_{j \in N_i} a_{ij}(q_i(t_k^i) - q_j(t_{k'}^j)), \\ \sum\limits_{i=1}^{n} w_i(0) = 0, \end{cases} \tag{8.4}$$

where $t \in [t_k^i, t_{k+1}^i)$, $t_0^i, t_1^i, \ldots, t_k^i, \ldots$ denote the trigger time sequence of agent i, and $t_{k'}^j = \max\{t_{k'}^j : t_{k'}^j < t\}$ denotes the latest event time of agent j before t_k^i.

With the proposed control protocol (8.4), the EL multi-agent systems (8.1) can be expressed as

$$\begin{cases} \dot{q}_i = p_i, \\ \dot{p}_i = -kp_i - \beta \sum\limits_{j \in N_i} a_{ij}(q_i(t_k^i) - q_j(t_{k'}^j)) - \nabla f_i(q_i) - w_i, \\ \dot{w}_i = \beta \sum\limits_{j \in N_i} a_{ij}(q_i(t_k^i) - q_j(t_{k'}^j)), \\ \sum\limits_{i=1}^{n} w_i(0) = 0. \end{cases} \tag{8.5}$$

Remark 8.2 $-\nabla f_i(q_i)$ in (8.5) is to optimize the local cost function of each agent, and $-w_i$ together with $\sum_{i=1}^{n} w_i(0) = 0$ and $\dot{w}_i = \beta \sum_{j \in N_i} a_{ij}(q_i(t_k^i) - q_j(t_{k'}^j))$ is to maintain the equilibrium point at the optimal point.

Remark 8.3 With the initial condition $\sum_{i=1}^{n} w_i(0) = 0$, we will show that the multi-agent system (8.1) has an equilibrium point at which the minimum value of the global objective function can be obtained. Besides, the design of w_i in (8.4) is independent of angular velocities, which is of benefit to reducing the communication cost among agents.

Proof By pre-multiplying $\dot{w}_i = \beta \sum_{j \in N_i} a_{ij}(q_i(t_k^i) - q_j(t_{k'}^j))$ with $(1_n^T \otimes I_m)$, we obtain that $\sum_{i=1}^{n} \dot{w}_i(t) = 0$. Thus

$$\sum_{i=1}^{n} \dot{w}_i(t) = \sum_{i=1}^{n} w_i(0) = 0, \forall t \geq 0. \tag{8.6}$$

Assume (q^*, p^*, w^*) are the equilibrium points of (8.5) and it follows that

$$\begin{cases} p^* = 0, \\ (L \otimes I_m)q^* = 0, \\ -w^* - \nabla f(q^*) = 0. \end{cases} \tag{8.7}$$

Since the graph is connected, it follows from $(L \otimes I_m)q^* = 0$ that $q^* = 1_n \otimes z_0$, $z_0 \in \mathbb{R}^m$, which means that consensus is achieved.

From (8.6), it is derived that $\sum_{i=1}^{n} w_i^* = 0$. Therefore, by pre-multiplying $-w^* - \nabla f(q^*) = 0$ with $(1_n^T \otimes I_m)$, we obtain that

$$\sum_{i=1}^{n} \nabla f_i(z_0) = 0, \tag{8.8}$$

which means that the global optimization is achieved. The equilibrium point z_0 is the optimal global minimizer z^*.

The position measurement errors for agent i are denoted as

$$e_q^i = q_i(t_k^i) - q_i(t). \tag{8.9}$$

Define the vector $e_q = [(e_q^1)^T, (e_q^2)^T, \ldots, (e_q^n)^T]^T$ and the triggering time sequence t_k^i, $k = 1, 2, \ldots$ are defined as

$$t_{k+1}^i = \inf\{t : t > t_k^i, f_e^i(q) > 0\}, \tag{8.10}$$

with

$$f_e^i(q) = \|e_q^i\| - \beta_1 \|\sum_{j \in N_i} a_{ij}(q_i(t_k^i) - q_j(t_{k'}^j))\| - \beta_2 e^{-\gamma t}, \tag{8.11}$$

where β_1, β_2 and γ are positive constants.

Remark 8.4 The purpose of the event-triggered scheme is to reduce the communication cost. Compared with the work in Liu et al. (2019b), the proposed event-triggering function (8.11) above avoids continuous communication.

Theorem 8.1 *Assume that Assumptions 8.1–8.3 hold. The distributed optimal consensus of EL system (8.1) under controller (8.4) with event-triggered function (8.11) can be exponentially achieved with a rate no less than* $e^{-\gamma t}$ *when parameters satisfy*

$$\begin{cases} k > \dfrac{3}{2}, \\ 0 < \delta < 1, \\ 0 < \beta_1 < \dfrac{1}{d + \sqrt{5}n}, \\ \omega - \omega_0 - 2\delta\theta^2 - \kappa_1 > 0, \end{cases}$$

where $\kappa_1 = \dfrac{5\beta_1^2(d^2+n^2)}{(1-\beta_1 d)^2 - 5\beta_1^2 n^2}$, $\omega_0 = \dfrac{\beta^2\lambda_{\max}^2(k+2)^2}{\delta} + \dfrac{\beta^2\lambda_{\max}^2}{\delta} + 2\beta\lambda_{\max}(k+1)$.

Proof Define $q = [q_1^T, \ldots, q_n^T]^T$, $p = [p_1^T, \ldots, p_n^T]^T$, $w = [w_1^T, \ldots, w_n^T]^T$, $\nabla f(q) = [\nabla f_1(q_1)^T, \ldots, \nabla f_n(q_n)^T]^T$. Thus, the system (8.5) can be rewritten as

$$\begin{cases} \dot{q} = p, \\ \dot{p} = -kp - \beta L(q + e_q) - \nabla f(q) - w, \\ \dot{w} = \beta L(q + e_q). \end{cases} \tag{8.12}$$

Further, introduce the following change of variables

$$\eta = q - q^*, \xi = p - p^* = p, \varpi = w - w^*,$$

and it is derived that

$$\begin{cases} \dot{\eta} = \xi, \\ \dot{\xi} = -k\xi - \beta L(\eta + e_q) - \varpi - h, \\ \dot{\varpi} = \beta L(\eta + e_q), \end{cases} \tag{8.13}$$

in which $h = \nabla f(\eta + q^*) - \nabla f(q^*)$.

Since $\mathcal{G}$ is connected, it is easy to know that $rank(L) = n - 1$ and the eigenvector corresponding to the eigenvalue 0 is spanned by $\mathbf{1}_n$. Set $Q = (r, U)$, where

$$r = \frac{\mathbf{1}_n}{\sqrt{n}}, \quad U \in \mathbb{R}^{n\times(n-1)}$$

satisfying

$$\mathbf{1}_n^T U = \mathbf{0}_{n-1}, \quad U^T U = I_{n-1}.$$

Thus it has $Q^T Q = I_n$, where Q is an orthogonal matrix. The following variables are introduced with the orthogonal transformations,

$$\hat{\eta} = (Q^T \otimes I_m)\eta, \hat{\xi} = (Q^T \otimes I_m)\xi, \hat{\varpi} = (Q^T \otimes I_m)\varpi,$$

$$\hat{e}_q = (Q^T \otimes I_m)e_q.$$

Then, the system (8.13) can be described as

$$\begin{cases} \dot{\hat{\eta}}_1 = \hat{\xi}_1, \\ \dot{\hat{\xi}}_1 = -k\hat{\xi}_1 - (r^T \otimes I_m)h, \\ \dot{\hat{\eta}}_{2:n} = \hat{\xi}_{2:n}, \\ \dot{\hat{\xi}}_{2:n} = -k\hat{\xi}_{2:n} - \beta(U^T LU \otimes I_m)(\hat{\eta}_{2:n} + \hat{e}_{q2:n}) \\ \qquad - \hat{\varpi}_{2:n} - (U^T \otimes I_m)h, \\ \dot{\hat{\varpi}}_1 = \mathbf{0}_m, \\ \dot{\hat{\varpi}}_{2:n} = \beta(U^T LU \otimes I_m)(\hat{\eta}_{2:n} + \hat{e}_{q2:n}), \end{cases} \tag{8.14}$$

where $\hat{\eta}_1, \hat{\xi}_1, \hat{\varpi}_1 \in \mathbb{R}^m$ and $\hat{\eta}_{2:n}, \hat{\xi}_{2:n}, \hat{\varpi}_{2:n} \in \mathbb{R}^{(n-1)\times m}$ with

$$\begin{aligned} \hat{\eta} &= [\hat{\eta}_1^T, \hat{\eta}_{2:n}^T]^T, \\ \hat{\xi} &= [\hat{\xi}_1^T, \hat{\xi}_{2:n}^T]^T, \\ \hat{\varpi} &= [\hat{\varpi}_1^T, \hat{\varpi}_{2:n}^T]^T, \\ \hat{e}_{q2:n} &= [(\hat{e}_q^2)^T, (\hat{e}_q^3)^T, \ldots, (\hat{e}_q^n)^T]^T. \end{aligned}$$

Set

$$V_0 = f(q) - f(q^*) - \nabla f(q^*)^T (q - q^*),$$

$$V_1 = \frac{1}{2}\hat{\xi}_1^T \hat{\xi}_1 + \hat{\xi}_1^T \hat{\eta}_1 + \frac{k}{2}\hat{\eta}_1^T \hat{\eta}_1,$$

$$V_2 = \frac{1}{2}\hat{\xi}_{2:n}^T \hat{\xi}_{2:n} + \hat{\xi}_{2:n}^T \hat{\eta}_{2:n} + \frac{k}{2}\hat{\eta}_{2:n}^T \hat{\eta}_{2:n},$$

$$V_3 = \frac{1}{2\beta}\hat{\varpi}_{2:n}^T ((U^T LU)^{-1} \otimes I_m)\hat{\varpi}_{2:n},$$

and

$$V_4 = \frac{1}{2}(\hat{\xi}_{2:n} + \hat{\varpi}_{2:n})^2,$$

$$V_5 = \frac{1}{2}(\hat{\eta}_{2:n} + \hat{\varpi}_{2:n})^2.$$

Define $\mu = [\hat{\xi}^T, \hat{\eta}^T, \hat{\varpi}_{2:n}^T]^T$ and choose a Lyapunov function for the whole system as

$$V = V_1 + V_2 + V_3 + V_4 + (k+2)V_5 = \mu^T \Sigma \mu, \tag{8.15}$$

with

$$\Sigma = \begin{pmatrix} \sigma_{11} & \sigma_{12} & \sigma_{13} \\ * & \sigma_{22} & \sigma_{23} \\ * & * & \sigma_{33} \end{pmatrix},$$

where $\sigma_{11} = \frac{1}{2}I_n + \frac{1}{2}\begin{pmatrix} 0 & 0_{n-1}^T \\ * & I_{n-1} \end{pmatrix}$, $\sigma_{22} = \frac{k}{2}I_n + \frac{k+2}{2}\begin{pmatrix} 0 & 0_{n-1}^T \\ * & I_{n-1} \end{pmatrix}$, $\sigma_{33} = \frac{1}{2\beta}(U^T LU)^{-1} + \frac{k+3}{2}I_{n-1}$, $\sigma_{12} = I_n$, $\sigma_{13} = \frac{1}{2}\begin{pmatrix} 0_{n-1}^T \\ I_{n-1} \end{pmatrix}$, and $\sigma_{23} = \frac{k+2}{2}\begin{pmatrix} 0_{n-1}^T \\ I_{n-1} \end{pmatrix}$.

The derivatives of V_1, V_2 and V_3 are respectively given as

$$\dot{V}_1 = -(k-1)\hat{\xi}_1^T\hat{\xi}_1 - (\hat{\xi}_1 + \hat{\eta}_1)^T(r^T \otimes I_m)h,$$

$$\begin{aligned}\dot{V}_2 = &-(k-1)\hat{\xi}_{2:n}^T\hat{\xi}_{2:n} - (\hat{\xi}_{2:n} + \hat{\eta}_{2:n})^T(U^T \otimes I_m)h \\ &- (\hat{\xi}_{2:n}^T + \hat{\eta}_{2:n}^T)\hat{\varpi}_{2:n} \\ &- \beta(\hat{\xi}_{2:n}^T + \hat{\eta}_{2:n}^T)(U^T LU \otimes I_m)(\hat{\eta}_{2:n} + \hat{e}_{q2:n}),\end{aligned}$$

$$\dot{V}_3 = \hat{\varpi}_{2:n}^T(\hat{\eta}_{2:n}^T + \hat{e}_{q2:n}),$$

$$\begin{aligned}\dot{V}_4 = &- k\hat{\xi}_{2:n}^T\hat{\xi}_{2:n} - \hat{\varpi}_{2:n}^T\hat{\varpi}_{2:n} \\ &- (\hat{\xi}_{2:n}^T + \hat{\varpi}_{2:n}^T)(U^T \otimes I_m)h - (k+1)\hat{\xi}_{2:n}^T\hat{\varpi}_{2:n},\end{aligned}$$

$$\begin{aligned}\dot{V}_5 =&(\hat{\eta}_{2:n} + \hat{\varpi}_{2:n})^T(\hat{\xi}_{2:n} + \beta L\tilde{\eta}_{2:n}) \\ =&\hat{\eta}_{2:n}^T\hat{\xi}_{2:n} + \hat{\varpi}_{2:n}^T\hat{\xi}_{2:n} + \beta\hat{\eta}_{2:n}^T(U^T LU \otimes I_m)(\hat{\eta}_{2:n} + \hat{e}_{q2:n}) \\ &+ \beta\hat{\varpi}_{2:n}(U^T LU \otimes I_m)(\hat{\eta}_{2:n} + \hat{e}_{q2:n}).\end{aligned}$$

Therefore,

$$\begin{aligned}\dot{V}_1 + \dot{V}_2 + \dot{V}_3 = &-(k-1)\hat{\xi}_1^T\hat{\xi}_1 - (k-1)\hat{\xi}_{2:n}^T\hat{\xi}_{2:n} \\ &- \beta(\hat{\xi}_{2:n}^T + \hat{\eta}_{2:n}^T)(U^T LU \otimes I_m)(\hat{\eta}_{2:n} + \hat{e}_{q2:n}) \\ &+ \hat{\varpi}_{2:n}^T\hat{e}_{q2:n} - \hat{\varpi}_{2:n}^T\hat{\xi}_{2:n} - (\xi + \eta)^T h,\end{aligned}$$

and

$$
\begin{aligned}
&\dot{V}_4 + (k+2)\dot{V}_5 \\
&= -k\hat{\xi}_{2:n}^T\hat{\xi}_{2:n} - \hat{\varpi}_{2:n}^T\hat{\varpi}_{2:n} - (\hat{\xi}_{2:n}^T + \hat{\varpi}_{2:n}^T)(U^T \otimes I_m)h \\
&\quad + (k+2)\hat{\eta}_{2:n}^T\hat{\xi}_{2:n} + \hat{\varpi}_{2:n}^T\hat{\xi}_{2:n} \\
&\quad + \beta(k+2)\hat{\eta}_{2:n}^T(U^T LU \otimes I_m)(\hat{\eta}_{2:n} + \hat{e}_{q2:n}) \\
&\quad + \beta(k+2)\hat{\varpi}_{2:n}(U^T LU \otimes I_m)(\hat{\eta}_{2:n} + \hat{e}_{q2:n}).
\end{aligned}
$$

From Assumption 8.2, it follows that

$$
-\eta^T h = -\sum_{i=1}^{n} \eta_i^T(\nabla f_i(\eta_i + q^*) - \nabla f_i(q^*)) \le -\omega\eta^T\eta. \tag{8.16}
$$

Therefore, we can have that

$$
\begin{aligned}
\dot{V} \le &-(k-1)\hat{\xi}_1^T\hat{\xi}_1 - (2k-1)\hat{\xi}_{2:n}^T\hat{\xi}_{2:n} - \hat{\varpi}_{2:n}^T\hat{\varpi}_{2:n} \\
&- \beta(\hat{\xi}_{2:n}^T + \hat{\eta}_{2:n}^T)(U^T LU \otimes I_m)(\hat{\eta}_{2:n} + \hat{e}_{q2:n}) \\
&- (\hat{\xi}_{2:n}^T + \hat{\varpi}_{2:n}^T)(U^T \otimes I_m)h - \xi^T h \\
&+ \hat{\varpi}_{2:n}^T\hat{e}_{q2:n} - \omega\eta^T\eta + (k+2)\hat{\eta}_{2:n}^T\hat{\xi}_{2:n} \\
&+ \beta(k+2)\hat{\eta}_{2:n}^T(U^T LU \otimes I_m)(\hat{\eta}_{2:n} + \hat{e}_{q2:n}) \\
&+ \beta(k+2)\hat{\varpi}_{2:n}^T(U^T LU \otimes I_m)(\hat{\eta}_{2:n} + \hat{e}_{q2:n}).
\end{aligned}
$$

For $\forall\delta > 0$, according to the inequality $x^T y \le \frac{\delta}{4}x^T x + \frac{1}{\delta}y^T y$, it follows that

$$
\begin{aligned}
\dot{V} \le &-(k-1)\hat{\xi}_1^T\hat{\xi}_1 - (2k-1)\hat{\xi}_{2:n}^T\hat{\xi}_{2:n} - \hat{\varpi}_{2:n}^T\hat{\varpi}_{2:n} \\
&- \beta(\hat{\xi}_{2:n}^T + \hat{\eta}_{2:n}^T)(U^T LU \otimes I_m)(\hat{\eta}_{2:n} + \hat{e}_{q2:n}) \\
&+ 2\delta\theta^2\eta^T\eta + \frac{\delta}{4}\hat{\varpi}_{2:n}^T\hat{\varpi}_{2:n} + \frac{\delta}{4}\hat{\xi}_{2:n}^T\hat{\xi}_{2:n} \\
&+ \frac{\delta}{4}\hat{\varpi}_{2:n}^T\hat{\varpi}_{2:n} + \frac{1}{\delta}\hat{e}_{q2:n}^T\hat{e}_{q2:n} + \frac{1}{2}\xi^T\xi + +\frac{\theta^2}{2}\eta^T\eta \\
&- \omega\eta^T\eta + \frac{\delta}{4}\hat{\xi}_{2:n}^T\hat{\xi}_{2:n} + \frac{(k+2)^2}{\delta}\hat{\eta}_{2:n}^T\hat{\eta}_{2:n} + \frac{\delta}{2}\hat{\varpi}_{2:n}^T\hat{\varpi}_{2:n} \\
&+ \beta(k+2)\hat{\eta}_{2:n}^T(U^T LU \otimes I_m)(\hat{\eta}_{2:n} + \hat{e}_{q2:n}) \\
&+ \frac{\beta^2(k+2)^2\lambda_{\max}^2}{\delta}\hat{\eta}_{2:n}^T\hat{\eta}_{2:n} + \frac{\beta^2(k+2)^2\lambda_{\max}^2}{\delta}\hat{e}_{q2:n}^T\hat{e}_{q2:n},
\end{aligned} \tag{8.17}
$$

where

$$\begin{aligned}
&-\beta(\hat{\xi}_{2:n}^T + \hat{\eta}_{2:n}^T)(U^T LU \otimes I_m)(\hat{\eta}_{2:n} + \hat{e}_{q2:n}) \\
&+\beta(k+2)\hat{\eta}_{2:n}^T(U^T LU \otimes I_m)(\hat{\eta}_{2:n} + \hat{e}_{q2:n}) \\
&\quad \le (\frac{\beta^2\lambda^2}{\delta} + \beta(k+1)\lambda + (k+1)^2)\hat{\eta}_{2:n}^T\hat{\eta}_{2:n} \\
&\quad + (\frac{\beta^2\lambda^2}{\delta} + \beta^2\lambda^2)\hat{e}_{q2:n}^T\hat{e}_{q2:n} + \frac{\delta}{2}\hat{\xi}_{2:n}^T\hat{\xi}_{2:n}.
\end{aligned} \tag{8.18}$$

Furthermore, it can be obtained that

$$\begin{aligned}
\dot{V} \le& -(k-3/2)\hat{\xi}_1^T\hat{\xi}_1 - (2k-\delta-3/2)\hat{\xi}_{2:n}^T\hat{\xi}_{2:n} \\
&- (1-\delta)\hat{\varpi}_{2:n}^T\hat{\varpi}_{2:n} - (\omega-\omega_0-2\delta\theta^2-\theta^2/2)\hat{\eta}^T\hat{\eta} \\
&+ (\omega_0 + \frac{1}{\delta})\hat{e}_{q2:n}^T\hat{e}_{q2:n},
\end{aligned} \tag{8.19}$$

where $\omega_0 = \dfrac{\beta^2\lambda_{\max}^2(k+2)^2}{\delta} + \dfrac{\beta^2\lambda_{\max}^2}{\delta} + 2\beta\lambda_{\max}(k+1)$.

From the triggering function, it can be obtained that

$$\begin{aligned}
\|e_q^i\| \le& \beta_1\|\sum_{j=1}^{n} a_{ij}(\eta_i(t_k^i) - \eta_j(t_{k'}^j))\| + \beta_2\exp(-\gamma t) \\
\le& \beta_1(\sum_{j=1}^{n} a_{ij}\|e_q^i\| + \sum_{j=1}^{n} a_{ij}\|e_q^j\| + \sum_{j=1}^{n} a_{ij}\|\eta_i\| \\
&+ \sum_{j=1}^{n} a_{ij}\|\eta_j\|) + \beta_2\exp(-\gamma t).
\end{aligned} \tag{8.20}$$

With the fact that $\sum_{j=1}^{n} a_{ij} = d$, the inequality (8.20) can be transformed as

$$\begin{aligned}
(1-\beta_1 d)^2\|e_q^i\|^2 \le& \Bigg(\beta_1\Bigg(\sum_{j=1}^{n} a_{ij}\|e_q^j\| + \sum_{j=1}^{n} a_{ij}\|\eta_i\| \\
&+ \sum_{j=1}^{n} a_{ij}\|\eta_j\|\Bigg) + \beta_2\exp(-\gamma t)\Bigg)^2.
\end{aligned} \tag{8.21}$$

Further, it can be derived that

$$
\begin{aligned}
&(1-\beta_1 d)^2\|e_q^i\|^2 \\
&\leq \Bigg(\beta_1^2\Big(\sum_{j=1}^n a_{ij}\|e_q^j\|\Big)^2+\beta_1^2 d^2\|\eta_i\|^2 \\
&+\beta_1^2\Big(\sum_{j=1}^n a_{ij}\|\eta_j\|\Big)^2+\beta_2^2\exp(-2\gamma t) \\
&+2\beta_1^2 d\|\eta_j\|\sum_{j=1}^n a_{ij}\|e_q^j\|+2\beta_1^2\sum_{j=1}^n a_{ij}\|e_q^j\|\sum_{j=1}^n a_{ij}\|\eta_j\| \\
&+2\beta_1\beta_2\exp(-\gamma t)\sum_{j=1}^n a_{ij}\|e_q^j\|+2\beta_1^2 d\|\eta_j\|\sum_{j=1}^n a_{ij}\|\eta_j\| \\
&+2\beta_1\beta_2 d\exp(-\gamma t)\|\eta_i\|+2\beta_1\beta_2\exp(-\gamma t)\sum_{j=1}^n a_{ij}\|\eta_j\|\Bigg).
\end{aligned}
\tag{8.22}
$$

In view of the fact that $\|e_q\|^2=\sum_{i=1}^n\|e_q^i\|^2$ and $2xy\leq x^2+y^2(\forall x,y\in\mathbb{R})$, it has

$$
\begin{aligned}
&\sum_{i=1}^n(1-\beta_1 d)^2\|e_q^i\|^2=(1-\beta_1 d)^2\|e_q\|^2 \\
&\leq\sum_{i=1}^n\Bigg(\beta_1^2\Big(\sum_{j=1}^n a_{ij}\|e_q^j\|\Big)^2+\beta_1^2 d^2\|\eta_i\|^2 \\
&\quad+\beta_1^2\Big(\sum_{j=1}^n a_{ij}\|\eta_j\|\Big)^2+\beta_2^2\exp(-2\gamma t) \\
&\quad+2\beta_1^2 d\|\eta_i\|\sum_{j=1}^n a_{ij}\|e_q^j\|+2\beta_1^2\sum_{j=1}^n a_{ij}\|e_q^j\|\sum_{j=1}^n a_{ij}\|\eta_j\| \\
&\quad+2\beta_1\beta_2\exp(-\gamma t)\sum_{j=1}^n a_{ij}\|e_q^j\|+2\beta_1^2 d\|\eta_j\|\sum_{j=1}^n a_{ij}\|\eta_j\| \\
&\quad+2\beta_1\beta_2 d\exp(-\gamma t)\|\eta_i\|+2\beta_1\beta_2\exp(-\gamma t)\sum_{j=1}^n a_{ij}\|\eta_j\|\Bigg) \\
&\leq\sum_{i=1}^n\Bigg(5\beta_1^2\Big(\sum_{j=1}^n a_{ij}\|e_q^j\|\Big)^2+5\beta_1^2 d^2\|\eta_i\|^2
\end{aligned}
$$

$$+ 5\beta_1^2\bigg(\sum_{j=1}^{n} a_{ij}\|\eta_j\|\bigg)^2 + 5\beta_2^2\exp(-2\gamma t)\bigg). \tag{8.23}$$

Furthermore, combining the fact that $\|\eta\|^2 = \sum_{i=1}^{n}\|\eta_i\|^2$, $(\sum_{j=1}^{n} a_{ij}\|e_q^j\|)^2 \leq n\sum_{i=1}^{n}\|e_q^j\|^2 = n\|e_q\|^2$, $(\sum_{j=1}^{n} a_{ij}\|\eta_j\|)^2 \leq n\sum_{i=1}^{n}\|\eta_j\|^2 = n\|\eta\|^2$, it is easily derived that

$$\|e_q\|^2 \leq \kappa_1\|\eta\|^2 + \kappa_2\exp(-2\gamma t), \tag{8.24}$$

where $\kappa_1 = \dfrac{5\beta_1^2(d^2+n^2)}{(1-\beta_1 d)^2 - 5\beta_1^2 n^2}$ and $\kappa_2 = \dfrac{5n\beta_2^2}{(1-\beta_1 d)^2 - 5\beta_1^2 n^2}$.

Furthermore, the derivative of V can be expressed as

$$\begin{aligned}\dot{V} &\leq -(k-3/2)\hat{\xi}_1^T\hat{\xi}_1 - (2k-\delta-3/2)\hat{\xi}_{2:n}^T\hat{\xi}_{2:n} \\ &\quad - (1-\delta)\hat{\varpi}_{2:n}^T\hat{\varpi}_{2:n} - (\omega-\omega_0-2\delta\theta^2-\theta^2/2-\kappa_1)\hat{\eta}^T\hat{\eta} \\ &\quad + \kappa_2\exp(-2\gamma t) \\ &\leq -\rho\mu^T\mu + \kappa_2\exp(-2\gamma t),\end{aligned} \tag{8.25}$$

where $\rho = \min\{k-3/2, 2k-\delta-3/2, 1-\delta, \omega-\omega_0-2\delta\theta^2-\theta^2/2-\kappa_1\}$.

From (8.15) and (8.25), it is derived that

$$\|\mu\|^2 \leq \frac{e^{-\frac{\rho}{\lambda_{\max}(\Sigma)}t}}{\lambda_{\min}(\Sigma)} + \frac{\kappa_2\lambda_{\max}(\Sigma)}{\lambda_{\min}(\Sigma)(\rho - 2\gamma\lambda_{\max}(\Sigma))}e^{-2\gamma t},$$

where $\lambda_{\max}(\Sigma)$ and $\lambda_{\min}(\Sigma)$ denote the maximum and minimum eigenvalues of the matrix Σ, respectively.

Select γ satisfying $0 < \gamma < \dfrac{\rho}{2\lambda_{\max}(\Sigma)}$, and it follows that

$$\|\mu\| \leq \sqrt{\frac{1}{\lambda_{\min}(\Sigma)} + \frac{\kappa_2\lambda_{\max}(\Sigma)}{\lambda_{\min}(\Sigma)(\rho - 2\gamma\lambda_{\max}(\Sigma))}}e^{-\gamma t}. \tag{8.26}$$

Thus, according to Theorem 8.1, when t tends to infinity, the optimal consensus can be achieved exponentially with a rate no less than $e^{-\gamma t}$.

Theorem 8.2 *Consider the distributed optimal control of the EL system (8.1) under event-triggered controllers (8.4). The Zeno behavior can be avoided with the triggering function (8.11).*

The following is to prove that the Zeno behavior can be ruled out.

Proof It is known that

$$\|\dot{e}_q^i\| = \|\xi_i\| \leq M\exp(-\gamma t). \tag{8.27}$$

Therefore, it follows that

$$\|e_q^i\| \leq -\frac{M}{\gamma}(\exp(-\gamma t) - \exp(-\gamma t_k^i)), \quad t \in [t_k^i, t_{k+1}^i). \tag{8.28}$$

When the next event happens, it follows that

$$\begin{aligned}\|e_q^i(t_{k+1}^i)\| &= \beta_1 \|\sum_{j=1}^n a_{ij}(q_i(t_k^i) - q_j(t_{k'}^j))\| \\ &+ \beta_2\exp(-\gamma t_{k+1}^i) \\ &\leq -\frac{M}{\gamma}(\exp(-\gamma t_{k+1}^i) - \exp(-\gamma t_k^i)).\end{aligned} \tag{8.29}$$

Define $\tau = t_{k+1}^i - t_k^i$, and it yields

$$\beta_2\exp(-\gamma\tau) \leq \frac{M}{\gamma}(1 - \exp(-\gamma\tau)). \tag{8.30}$$

Furthermore, it follows that

$$\tau = \frac{1}{\gamma}\ln(1 + \frac{\gamma\beta_2}{M}) > 0, \tag{8.31}$$

which means that the Zeno behavior is ruled out for all the agents i.

8.4 Event-Triggered Optimal Consensus with Unavailable Model Parameters for EL Systems

In subsection III.A, the model parameters of the EL systems are assumed to be available, which can simplify the system (8.1) to a double-integrator system. However, when the model parameters of EL systems are unavailable, the algorithms in subsection III.A cannot be used directly. In this section, a reference model and tracking controllers are introduced to solve this problem.

From the property of the EL systems, the system (8.1) can be rewritten as

$$M_i(q_i)\ddot{\hat{q}}_i + C_i(q_i, \dot{q}_i)\dot{\hat{q}}_i + G_i(q_i) = \Omega_i\varphi_i, \tag{8.32}$$

where $\Omega_i = \Omega_i(q_i, \dot{q}_i, \dot{\hat{q}}_i, \ddot{\hat{q}}_i) \in \mathbb{R}^{m\times p}$ is a known matrix and $\varphi_i \in \mathbb{R}^p$ is an unknown constant vector. Denote $\hat{q}_i = y_i - (q_i - x_i)$ and $e_i = \dot{q}_i - \dot{\hat{q}}_i$.

A virtual double-integrator system is constructed by

$$\begin{cases} \dot{x}_i = y_i, \\ \dot{y}_i = u_i. \end{cases} \tag{8.33}$$

Remark 8.5 The virtual system above can be viewed as a reference model. Hence, the objective of this subsection is to design tracking controllers so that the EL multi-agent system (8.1) can follow the desired trajectories produced by the virtual system (8.33).

Therefore, we define the following optimal controllers

$$\begin{cases} \dot{x}_i = y_i, \\ \dot{y}_i = u_i, \\ u_i = -ky_i - \beta\sum_{j=1}^{n} a_{ij}(x_i(t_k^i) - x_j(t_{k'}^j)) - \nabla f_i(x_i) - w_i, \\ \dot{w}_i = \sum_{j=1}^{n} a_{ij}(x_i(t_k^i) - x_j(t_{k'}^j)), \\ \sum_{i=1}^{n} w_i(0) = 0, \end{cases} \tag{8.34}$$

and tracking controllers

$$\begin{cases} \tau_i = -\Xi_i e_i + \Omega_i \psi_i, \\ \dot{\psi}_i = -\Omega_i^T e_i, \\ e_i = (\dot{q}_i - y_i) + (q_i - x_i), \end{cases} \tag{8.35}$$

where $\Xi_i \in \mathbb{R}^{m\times m}$ is positive-definite and ψ_i is the estimate of the unknown parameter φ_i.

To introduce the event-triggering mechanism, define the measurement errors as

$$e_x^i = x_i(t_k^i) - x_i(t). \tag{8.36}$$

Define the vector $e_x = [(e_x^1)^T, (e_x^2)^T, \ldots, (e_x^N)^T]^T$ and the triggering time sequence $t_k^i, k = 1, 2, \ldots$ are defined as

$$t_{k+1}^i = \inf\{t : t > t_k^i, f^i(x) > 0\}, \tag{8.37}$$

with

$$f^i(x) = \|e_x^i\| - \beta_3 \| \sum_{j \in N_i} a_{ij}(x_i(t_k^i) - x_j(t_{k'}^j))\| - \beta_4 e^{-\tilde{\gamma}t}, \tag{8.38}$$

where $t \in [t_k^i, t_{k+1}^i)$, $t_0^i, t_1^i, \ldots, t_k^i, \ldots$ denote the trigger time sequence of agent i, $t_{k'}^j = \max\{t_{k'}^j : t_{k'}^j < t\}$ denotes the latest event time of agent j before t_k^i, and β_3, β_4, $\tilde{\gamma}$ are positive constants.

Theorem 8.3 *Assume that Assumptions 8.1–8.3 hold. The distributed optimal consensus of the EL system (8.1) with event-triggered controllers (8.34), (8.35) and triggering function (8.38) can be exponentially achieved with a rate no less than $e^{-\tilde{\gamma}t}$ when parameters satisfy*

$$\begin{cases} k > \dfrac{3}{2}, \\ 0 < \delta < 1, \\ 0 < \beta_3 < \dfrac{1}{d + \sqrt{5}n}, \\ \omega - \omega_0 - 2\delta\theta^2 - \kappa_3 > 0, \end{cases}$$

where $\kappa_3 = \dfrac{5\beta_3^2(d^2+n^2)}{(1-\beta_3 d)^2 - 5\beta_3^2 n^2}$, $\omega_0 = \dfrac{\beta^4\lambda_{\max}^2(k+2)^2}{\delta} + \dfrac{\beta^2\lambda_{\max}^2}{\delta} + 2\beta\lambda_{\max}(k+1)$. *Moreover, the proposed algorithms (8.34), (8.35) and (8.38) exclude the Zeno behavior.*

Proof Under the controller (8.34), it follows that

$$M_i(q_i)\ddot{q}_i + C_i(q_i, \dot{q}_i)\dot{q}_i + G_i(q_i) = -\Xi_i e_i + \Omega_i \psi_i. \tag{8.39}$$

Moreover, with the definition of tracking error e_i, it is derived that

$$M_i(q_i)\dot{e}_i + C_i(q_i, \dot{q}_i)e_i = -\Xi_i e_i + \Omega_i(\psi_i - \varphi_i). \tag{8.40}$$

Define $e = [e_1^T, e_2^T, \ldots, e_n^T]^T$ and construct the Lyapunov function

$$V' = e^T M(q)e + (\psi - \varphi)^T(\psi - \varphi). \tag{8.41}$$

Therefore, the derivation of the V is expressed as

$$\begin{aligned} \dot{V}' &= 2e^T M(q)\dot{e} + e^T \dot{M}(q)e + 2(\psi - \varphi)^T \dot{\psi} \\ &= 2e^T M(q)\dot{e} + 2e^T C(q, \dot{q})e - 2(\psi - \varphi)^T \Omega^T e \\ &= 2e^T(-\Xi e + \Omega(\psi - \varphi)) - 2(\psi - \varphi)^T \Omega^T e \\ &= -2e^T \Xi e. \end{aligned} \tag{8.42}$$

By Barbalat's Lemma, it yields $\lim_{t\to\infty} e(t) = 0_m$. It is known that

$$\dot{q}_i - \dot{x}_i = -(q_i - x_i) + e_i. \tag{8.43}$$

Therefore, it follows that $\lim_{t\to\infty}(q_i - x_i) = 0_m$ and $\lim_{t\to\infty}(\dot{q}_i - \dot{x}_i) = 0_m$, which implies

$$\begin{aligned} \lim_{t\to\infty} q_i(t) &= \lim_{t\to\infty} x_i(t), \\ \lim_{t\to\infty} \dot{q}_i(t) &= \lim_{t\to\infty} y_i(t). \end{aligned} \tag{8.44}$$

Therefore, it is easy to know that the EL system (8.1) can track the virtual double-integrator system (8.33).

The equilibrium point of the virtual system (8.34) is denoted by (x^*, y^*, w^*) satisfying

$$\begin{cases} y^* = 0, \\ (L \otimes I_m)x^* = 0, \\ w^* + \nabla f(x^*) = 0. \end{cases} \tag{8.45}$$

Since $\mathcal{G}$ is connected, we have $rank(L) = n - 1$ and the eigenvector corresponding to eigenvalue 0 is spanned by $\mathbf{1}_n$. Introduce the following variables

$$\eta = x - x^*, \xi = y - y^* = y, \varpi = w - w^*,$$

and the system (8.34) can be expressed as

$$\begin{cases} \dot{\eta} = \xi, \\ \dot{\xi} = -k\xi - \beta L(\eta + e_x) - \varpi - h, \\ \dot{\varpi} = \beta L(\eta + e_x), \end{cases} \tag{8.46}$$

in which $h = \nabla f(\eta + x^*) - \nabla f(x^*)$.

Introduce an orthogonal matrix $Q = (r, U)$, in which

$$r = \frac{\mathbf{1}_n}{\sqrt{n}}, U \in \mathbb{R}^{n\times(n-1)}$$

satisfying

$$\mathbf{1}_n^T U = \mathbf{0}_{n-1}, U^T U = I_{n-1}.$$

It is easy to derive that $Q^T Q = I_n$. Furthermore, introduce the following variables

$$\hat{\eta} = (Q^T \otimes I_m)\eta, \quad \hat{\xi} = (Q^T \otimes I_m)\xi,$$

$$\hat{\varpi} = (Q^T \otimes I_m)\varpi, \quad \hat{e}_x = (Q^T \otimes I_m)e_x,$$

and the system (8.46) can be transformed as

$$\begin{cases} \dot{\hat{\eta}}_1 = \hat{\xi}_1, \\ \dot{\hat{\xi}}_1 = -k\hat{\xi}_1 - (r^T \otimes I_m)h, \\ \dot{\hat{\eta}}_{2:n} = \hat{\xi}_{2:n}, \\ \dot{\hat{\xi}}_{2:n} = -k\hat{\xi}_{2:n} - \beta(U^T LU \otimes I_m)(\hat{\eta}_{2:n} + \hat{e}_{x2:n}) \\ \quad - \hat{\varpi}_{2:n} - (U^T \otimes I_m)h, \\ \dot{\hat{\varpi}}_1 = \mathbf{0}_m, \\ \dot{\hat{\varpi}}_{2:n} = \beta(U^T LU \otimes I_m)(\hat{\eta}_{2:n} + \hat{e}_{x2:n}), \end{cases} \tag{8.47}$$

where $\hat{\eta}_1, \hat{\xi}_1, \hat{\varpi}_1 \in \mathbb{R}^m$ and $\hat{\eta}_{2:n}, \hat{\xi}_{2:n}, \hat{\varpi}_{2:n} \in \mathbb{R}^{(n-1)\times m}$ with

$$\begin{aligned} \hat{\eta} &= [\hat{\eta}_1^T, \hat{\eta}_{2:n}^T]^T, \\ \hat{\xi} &= [\hat{\xi}_1^T, \hat{\xi}_{2:n}^T]^T, \\ \hat{\varpi} &= [\hat{\varpi}_1^T, \hat{\varpi}_{2:n}^T]^T, \\ \hat{e}_{x2:n} &= [(\hat{e}_x^2)^T, (\hat{e}_x^3)^T, \ldots, (\hat{e}_x^n)^T]^T. \end{aligned}$$

Set

$$V_0 = f(x) - f(x^*) - \nabla f(x^*)^T (x - x^*)$$

$$V_1 = \frac{1}{2}\hat{\xi}_1^T \hat{\xi}_1 + \hat{\xi}_1^T \hat{\eta}_1 + \frac{k}{2}\hat{\eta}_1^T \hat{\eta}_1,$$

$$V_2 = \frac{1}{2}\hat{\xi}_{2:n}^T \hat{\xi}_{2:n} + \hat{\xi}_{2:n}^T \hat{\eta}_{2:n} + \frac{k}{2}\hat{\eta}_{2:n}^T \hat{\eta}_{2:n},$$

$$V_3 = \frac{1}{2\beta}\hat{\varpi}_{2:n}^T ((U^T LU)^{-1} \otimes I_m)\hat{\varpi}_{2:n}$$

and

$$V_4 = \frac{1}{2}(\hat{\xi}_{2:n} + \hat{\varpi}_{2:n})^2,$$

$$V_5 = \frac{1}{2}(\hat{\eta}_{2:n} + \hat{\varpi}_{2:n})^2.$$

Define $\mu = [\hat{\xi}^T, \hat{\eta}^T, \hat{\varpi}_{2:n}^T]^T$ and construct a candidate Lyapunov function for the whole system

$$V = V_1 + V_2 + V_3 + V_4 + (k+2)V_5 = \mu^T \Sigma \mu. \tag{8.48}$$

Similar to the procedure in the proof of Theorem 1, it is obtained that

$$\begin{aligned}\dot{V} \leq & -(k-3/2)\hat{\xi}_1^T\hat{\xi}_1 - (2k-\delta-3/2)\hat{\xi}_{2:n}^T\hat{\xi}_{2:n} \\ & -(1-\delta)\hat{\varpi}_{2:n}^T\hat{\varpi}_{2:n} - (\omega-\omega_0-2\delta\theta^2-\theta^2/2)\hat{\eta}^T\hat{\eta} \\ & +(\omega_0+\frac{1}{\delta})\hat{e}_{x2:n}^T\hat{e}_{x2:n},\end{aligned} \tag{8.49}$$

where $\omega_0 = \dfrac{\beta^2\lambda_{\max}^2(k+2)^2}{\delta} + \dfrac{\beta^2\lambda_{\max}^2}{\delta} + 2\beta\lambda_{\max}(k+1)$.

From the triggering function, it follows that

$$\begin{aligned}\|e_x^i\| \leq & \beta_3\|\sum_{j=1}^n a_{ij}(\eta_i(t_k^i) - \eta_j(t_{k'}^j))\| + \beta_4\exp(-\tilde{\gamma}t) \\ \leq & \beta_3\Big(\sum_{j=1}^n a_{ij}\|e_x^i\| + \sum_{j=1}^n a_{ij}\|e_x^j\| + \sum_{j=1}^n a_{ij}\|\eta_i\| \\ & + \sum_{j=1}^n a_{ij}\|\eta_j\|\Big) + \beta_4\exp(-\tilde{\gamma}t).\end{aligned} \tag{8.50}$$

Then, from the fact that $\|e_x\|^2 = \sum_{i=1}^n \|e_x^i\|^2$, $\|\eta\|^2 = \sum_{i=1}^n \|\eta_i\|^2$, $(\sum_{i=1}^n \|\eta_i\|)^2 \leq n\sum_{i=1}^n \|\eta_i\|^2 = n\|\eta\|^2$, $2xy \leq x^2 + y^2 (\forall x, y \in \mathbb{R})$, it is derived that

$$\|e_x\|^2 \leq \kappa_3\|\eta\|^2 + \kappa_4\exp(-2\tilde{\gamma}t), \tag{8.51}$$

where $\kappa_3 = \dfrac{5\beta_3^2(d^2+n^2)}{(1-\beta_3 d)^2 - 5\beta_3^2 n^2}$ and $\kappa_4 = \dfrac{5n\beta_4^2}{(1-\beta_3 d)^2 - 5\beta_3^2 n^2}$.

Furthermore, the derivative of V can be expressed as

$$\begin{aligned}\dot{V} \leq & -(k-3/2)\hat{\xi}_1^T\hat{\xi}_1 - (2k-\delta-3/2)\hat{\xi}_{2:n}^T\hat{\xi}_{2:n} \\ & -(1-\delta)\hat{\varpi}_{2:n}^T\hat{\varpi}_{2:n} - (\omega-\omega_0-2\delta\theta^2-\theta^2/2-\kappa_3)\hat{\eta}^T\hat{\eta} \\ & +\kappa_4\exp(-2\tilde{\gamma}t) \\ \leq & -\tilde{\rho}\mu^T\mu + \kappa_4\exp(-2\tilde{\gamma}t),\end{aligned} \tag{8.52}$$

where $\tilde{\rho} = \min\{k-3/2, 2k-\delta-3/2, 1-\delta, \omega-\omega_0-2\delta\theta^2-\theta^2/2-\kappa_3\}$.

From (8.48) and (8.52), it is derived that

$$\|\mu\|^2 \leq \frac{e^{-\frac{\tilde{\rho}}{\lambda_{\max}(\Sigma)}t}}{\lambda_{\min}(\Sigma)} + \frac{\kappa_4\lambda_{\max}(\Sigma)}{\lambda_{\min}(\Sigma)(\tilde{\rho}-2\tilde{\gamma}\lambda_{\max}(\Sigma))}e^{-2\tilde{\gamma}t}.$$

Select $\tilde{\gamma}$ satisfying $0 < \tilde{\gamma} < \dfrac{\tilde{\rho}}{2\lambda_{\max}(\Sigma)}$, and it yields

$$\|\mu\| \leq \sqrt{\frac{1}{\lambda_{\min}(\Sigma)} + \frac{\kappa_4 \lambda_{\max}(\Sigma)}{\lambda_{\min}(\Sigma)(\tilde{\rho} - 2\gamma\lambda_{\max}(\Sigma))}} e^{-\tilde{\gamma}t}. \tag{8.53}$$

Thus, when t tends to infinity, the optimal consensus can be achieved exponentially with a rate no less than $e^{-\tilde{\gamma}t}$. The proof of excluding Zeno behavior is similar to Theorem 8.2, so it is omitted in this subsection to avoid repetition.

Remark 8.6 The proposed algorithm in this chapter extends the results in Zhang et al. (2017) with event-triggered scheme to reduce communication cost. Different from the result of semi-global convergence in Deng and Yiguang (2016), the proposed algorithms can guarantee the EL systems to achieve global convergence exponentially.

8.5 Numerical Simulation

In this section, a numerical simulation is given to demonstrate the proposed theoretical algorithms and show the effectiveness of the distributed proximal continuous control algorithms.

Give six networked two-link revolute joint arms, which are modeled by Euler-Lagrange equations (Liu et al. 2019b; Mei et al. 2011; Spong and Hutchinson 2006). The communication topology is shown in the dotted box (see Fig. 8.1), where q_i denotes the joint angle. The system dynamics are given by

$$\begin{bmatrix} M_{11,i} & M_{12,i} \\ M_{21,i} & M_{22,i} \end{bmatrix} \begin{bmatrix} \ddot{q}_{i1} \\ \ddot{q}_{i2} \end{bmatrix} + \begin{bmatrix} C_{11,i} & C_{12,i} \\ C_{21,i} & C_{22,i} \end{bmatrix} \begin{bmatrix} \dot{q}_{i1} \\ \dot{q}_{i2} \end{bmatrix} = \begin{bmatrix} \tau_{i1} \\ \tau_{i2} \end{bmatrix},$$

where $i = 1, 2, \ldots, 6$, $M_{11,i} = b_{i1} + 2b_{i2}\cos(q_{i2})$, $M_{12,i} = b_{i3} + b_{i2}\cos(q_{i2})$, $M_{21,i} = b_{i3} + 2b_{i2}\cos(q_{i2})$, $M_{22,i} = b_{i3}$, $C_{11,i} = -b_{i2}\sin(q_{i2})\dot{q}_{i2}$, $C_{12,i} = -b_{i2}\sin(q_{i2})(\dot{q}_{i1} + \dot{q}_{i2})$, $C_{21,i} = b_{i2}\sin(q_{i2})\dot{q}_{i1}$, and $C_{22,i} = 0$. Referring to Liu et al. (2019b), Mei et al. (2011) and Spong and Hutchinson (2006), set $b_{i1} = 1.301$, $b_{i2} = 0.256$, $b_{i3} = 0.096$, and the initial states of the agents are given randomly. Denote the local cost function of each agent i by $f_i(q_i) = \|q_i - R_i\|^2$, where R_i is the reference points of each agent, $q_i = [q_{i1}, q_{i2}]^T$ and $R_1 = [3, 2]^T$, $R_2 = [4, -2]^T$, $R_3 = [-2, -4]^T$, $R_4 = [-6, -1]^T$, $R_5 = [-3, 2]^T$, $R_6 = [-1, 2]^T$. The cost function shows that the objective of each agent is to stay closer to its reference points R_i, and it is easy to derive that Assumptions 2 and 3 are satisfied. The parameters in (8.34) and (8.38) are designed as $k = 5$, $\beta = 1$, $\beta_3 = 0.002$, $\beta_4 = 0.004$, $\tilde{\gamma} = 0.2$.

Figures 8.2, 8.3 and 8.4 describe the trajectories of the angles and angular velocities of the six agents with algorithms (8.34) and (8.35). In Fig. 8.2, q^* is the optimal

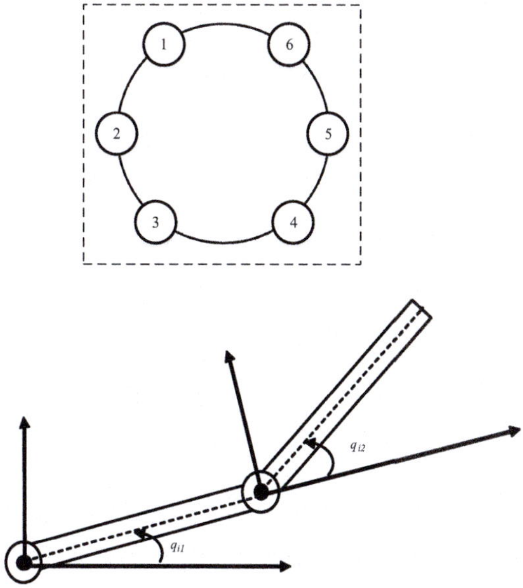

Fig. 8.1 Two-link manipulator with generalized coordinates $q_i = [q_{i1}, q_{i2}]^T$ $(i = 1, 2, \ldots, 6)$

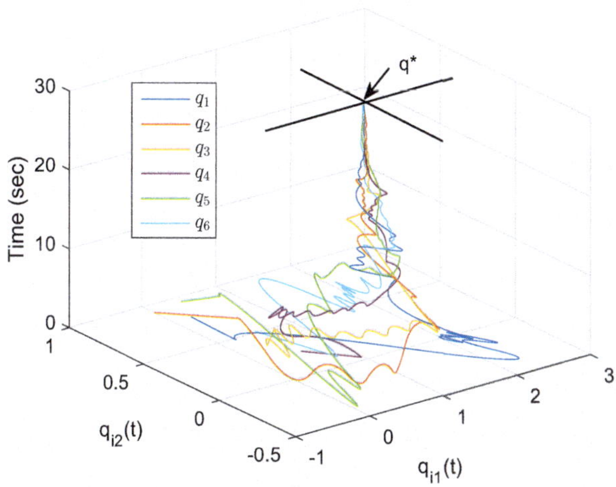

Fig. 8.2 The trajectories of q_i $(i = 1, 2, \ldots, 6)$ with algorithms shown in (8.34) and (8.35)

angle for all agents. Figures 8.3 and 8.4 show that the angular velocities of the agents reach zero when the optimal angle is reached. Figures 8.5 and 8.6 denote the event-triggered control curves u_{i1} and $u_{i2}, i = 1, \ldots, 6$ in algorithm (8.34). When the optimal angle is reached, the control of the virtual system becomes zero. Figures 8.7 and 8.8 describe the measurement errors with algorithm (8.34), which decrease over time.

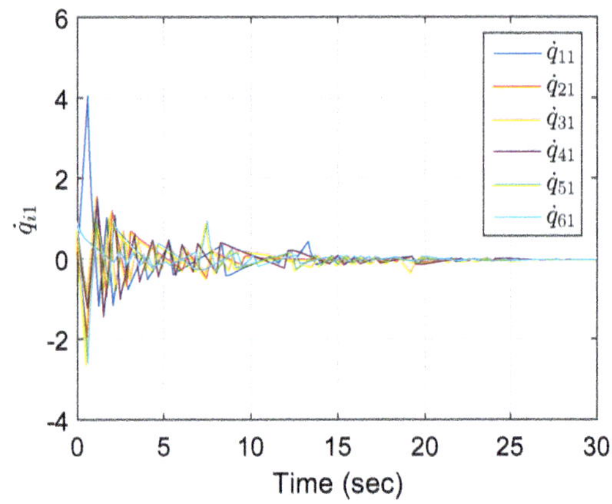

Fig. 8.3 The trajectories of $\dot{q}_{i1}$ $(i = 1, 2, \ldots, 6)$ with algorithms shown in (8.34) and (8.35)

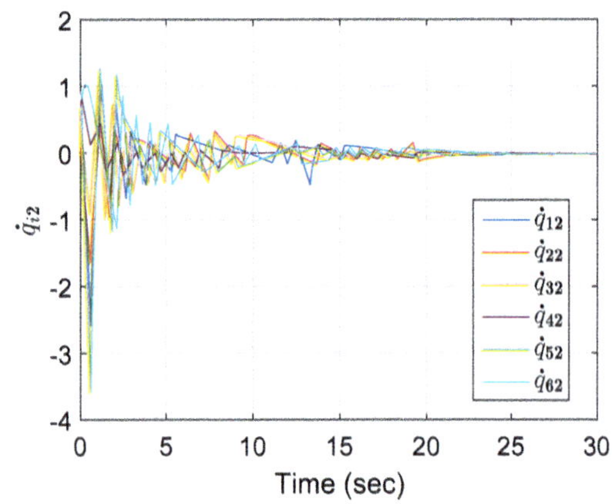

Fig. 8.4 The trajectories of $\dot{q}_{i2}$ $(i = 1, 2, \ldots, 6)$ with algorithms shown in (8.34) and (8.35)

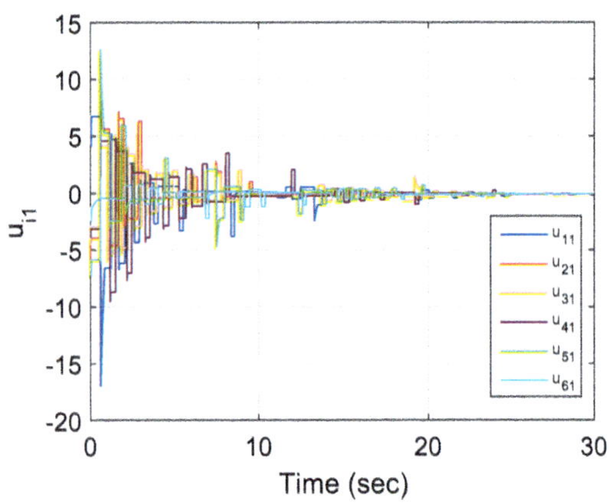

Fig. 8.5 The event-triggered control curves of u_{i1} $(i = 1, 2, \ldots, 6)$ with algorithm shown in (8.34)

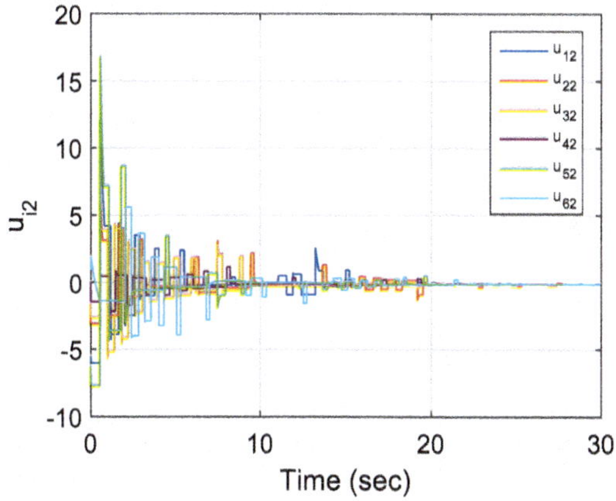

Fig. 8.6 The event-triggered control curves of u_{i2} $(i = 1, 2, \ldots, 6)$ with algorithm shown in (8.34)

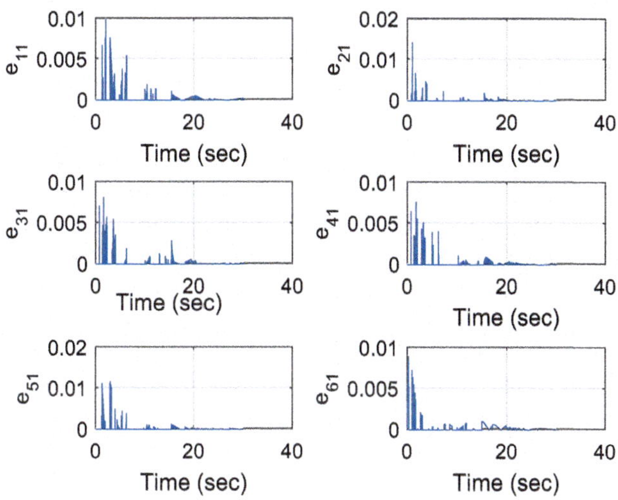

Fig. 8.7 The measurement errors e_{i1} $(i = 1, 2, \ldots, 6)$

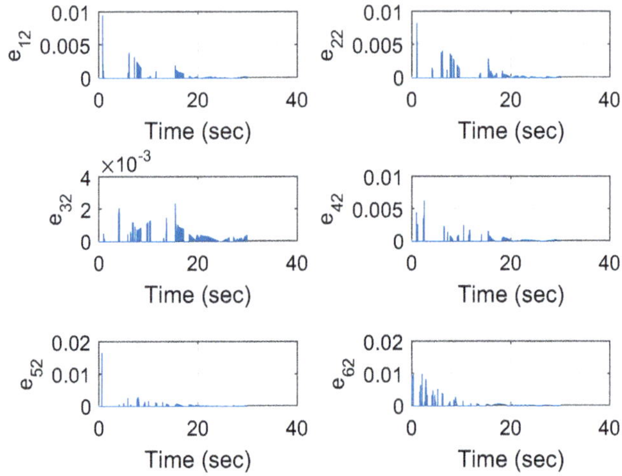

Fig. 8.8 The measurement errors e_{i2} $(i = 1, 2, \ldots, 6)$

8.6 Conclusion

In this chapter, the distributed convex optimization based on the event-triggered scheme is studied for Euler-Lagrange multi-agent systems when the model parameters are available and unavailable, respectively. With the proposed algorithms, the global cost function can be minimized in a distributed manner while achieving consensus with lower communication cost. Finally, theoretical proof is provided to show that the proposed algorithms can ensure the system to achieve global convergence exponentially. In the future, the optimal algorithms based on the event-triggered scheme will be developed for Euler-Lagrange systems with uncertainties under switching or directed topologies.

References

Deng, Z. and Yiguang, H. (2016). "Multi-agent optimization design for autonomous lagrangian systems," *Unmanned Systems***1**, 4, pp. 5–13.

Deng, Z. H., Wang, X. H., and Hong, Y. G. (2017). "Distributed optimisation design with triggers for disturbed continuous-time multi-agent systems," *IET Control Theory and Applications***11**, 2, pp. 282–290.

Dimarogonas, D. V., Frazzoli, E., and Johansson, K. H. (2012). "Distributed event-triggered control for multi-agent systems," *IEEE Transactions on Automatic Control***57**, 5, pp. 1291–1297.

Kia, S. S., Cortes, J., and Martinez, S. (2015). "Distributed convex optimization via continuous-time coordination algorithms with discrete-time communication," *Automatica***55**, pp. 254–264.

Li, H. Q., Liu, S., Soh, Y. C., and Xie, L. H. (2018). "Event-triggered communication and data rate constraint for distributed optimization of multiagent systems," *IEEE Transactions on Systems Man Cybernetics-Systems***48**, 11, pp. 1908–1919.

Liu, J., Yu, Y., Sun, J., and Sun, C. Y. (2018). "Distributed event-triggered fixed-time consensus for leader-follower multiagent systems with nonlinear dynamics and uncertain disturbances," *International Journal of Robust and Nonlinear Control* **28**, 11, pp. 3543–3559.

Liu, J., Zhang, Y., Yu, Y. and Sun, C. (2019). "Fixed-time event-triggered consensus for nonlinear multiagent systems without continuous communications," *IEEE Transactions on Systems, Man, and Cybernetics: Systems* **49**, 11, pp. 2221–2229.

Liu, J., Zhang, Y. L., Sun, C. Y., and Yu, Y. (2019a). "Fixed-time consensus of multi-agent systems with input delay and uncertain disturbances via event-triggered control," *Information Sciences* **480**, pp. 261–272.

Liu, Q. C., Ye, M. B., Qin, J. H., and Yu, C. B. (2019b). "Event-triggered algorithms for leader-follower consensus of networked euler-lagrange agents," *IEEE Transactions on Systems Man Cybernetics-Systems* **49**, 7, pp. 1435–1447.

Mei, J., Ren, W., and Ma, G. F. (2011). "Distributed coordinated tracking with a dynamic leader for multiple euler-lagrange systems," *IEEE Transactions on Automatic Control* **56**, 6, pp. 1415–1421.

Qiu, Z. and Yiguang, H. (2016). "Optimal consensus of euler-lagrangian systems with kinematic constraints," *IFAC-PapersOnLine* **22**, 49, pp. 327–332.

Shi, X. S., Wang, Y. D., Song, S. M., and Yan, G. F. (2018). "Distributed optimisation for resource allocation with event-triggered communication over general directed topology," *International Journal of Systems Science* **49**, 6, pp. 1119–1130.

Spong, M. W. and Hutchinson (2006). *Robot modeling and control* (John Wiley and Sons).

Tran, N. T., Wang, Y. W., Liu, X. K., Xiao, J. W., and Lei, Y. (2019). "Distributed optimization problem for second-order multi-agent systems with event-triggered and time-triggered communication," *Journal of the Franklin Institute-Engineering and Applied Mathematics* **356**, 17, pp. 10196–10215.

Wang, X. F., Deng, Z. H., Ma, S., and Du, X. (2017). "Event-triggered design for multi-agent optimal consensus of euler-lagrangian systems," *kybernetika* **53**, 1, pp. 179–194.

Wu, Z. G., Xu, Y., Lu, R. Q., Wu, Y. Q., and Huang, T. W. (2018). "Event-triggered control for consensus of multiagent systems with fixed/switching topologies," *IEEE Transactions on Systems Man Cybernetics-Systems* **48**, 10, pp. 1736–1746.

Yi, X., Yao, L., Yang, T., George, J., and Johansson, K. H. (2018). "Distributed optimization for second-order multi-agent systems with dynamic event-triggered communication," in *2018 IEEE Conference on Decision and Control (CDC)* (IEEE), pp. 3397–3402.

Zhang, Y. Q., Deng, Z. H., and Hong, Y. G. (2017). "Distributed optimal coordination for multiple heterogeneous euler-lagrangian systems," *Automatica* **79**, pp. 207–213.

Zhang, Y., Su, X. H., Liu, Z., and Chen, C. (2021). "Event-triggered adaptive fuzzy tracking control with guaranteed transient performance for mimo nonlinear uncertain systems," *IEEE Transactions on Cybernetics* **51**, 2, pp. 736–749.

Zhao, Z. Y., Chen, G., and Dai, M. X. (2018). "Distributed event-triggered scheme for a convex optimization problem in multi-agent systems," *Neurocomputing* **284**, pp. 90–98.

The manufacturer's authorised representative in the EU is Springer Nature Customer Service Centre GmbH, Europaplatz 3, 69115 Heidelberg, Germany. If you have any concerns regarding our products, please contact ProductSafety@springernature.com

Printed and bound by CPI Group (UK) Ltd, Croydon, CR0 4YY
30/09/2026
02225308-0007